Macromolecular Symposia 216

Contributions from 8th Pacific Polymer Conference

Bangkok, Thailand
November 24–27, 2003

Symposium Editors:
P. Supaphol, S. Tantayanon, Bangkok, Thailand

pp. 1–306 · September 2004
ISBN 3-527-31049-5

Macromolecular Symposia publishes lectures given at international symposia and is issued irregularly, with normally 14 volumes published per year. For each symposium volume, an Editor is appointed. The articles are peer-reviewed. The journal is produced by photo-offset lithography directly from the authors' typescripts.
Further information for authors can be found at http://www.ms-journal.de
Suggestions or proposals for conferences or symposia to be covered in this series should also be sent to the Editorial office (E-mail: macro-symp@wiley-vch.de).

Macromolecular Symposia:
Annual subscription rates 2005
Macromolecular Full Package: including Macromolecular Chemistry & Physics (24 issues), Macromolecular Rapid Communications (24), Macromolecular Bioscience (12), Macromolecular Theory & Simulations (9), Macromolecular Materials and Engineering (12), Macromolecular Symposia (14):

Europe	Euro	7.088 / 7.797
Switzerland	Sfr	12.448 / 13.693
All other areas	US$	8.898 / 9.788

print only **or** electronic only / print **and** electronic

Postage and handling charges included. All Wiley-VCH prices are exclusive of VAT. Prices are subject to change.

Single issues and back copies are available. Please ask for details at: service@wiley-vch.de

Orders may be placed through your bookseller or directly at the publishers:
WILEY-VCH Verlag GmbH & Co. KGaA, P. O. Box 10 11 61, 69451 Weinheim, Germany,
Tel. +49 (0) 62 01/6 06-400, Fax +49 (0) 62 01/60 61 84, E-mail: service@wiley-vch.de
For copyright permission please contact Claudia Jerke at:
Fax: +49 (0) 62 01/6 06-332, E-mail: cjerke@wiley-vch.de

For USA and Canada: Macromolecular Symposia (ISSN 1022-1360) is published with 14 volumes per year by WILEY-VCH Verlag GmbH & Co. KGaA, Boschstr. 12, 69451 Weinheim, Germany. Air freight and mailing in the USA by Publications Expediting Inc., 200 Meacham Ave., Elmont, NY 11003, USA. Application to mail at Periodicals Postage rate is pending at Jamaica, NY 11431, USA. POSTMASTER please send address changes to: Macromolecular Symposia, c/o Wiley-VCH, III River Street, Hoboken, NJ 07030, USA.

Macromolecular Symposia

Articles published on the web will appear several weeks before the print edition. They are available through:

www.ms-journal.de

www.interscience.wiley.com

8th Pacific Polymer Conference
Bangkok (Thailand), 2003

Preface
P. Supaphol, S. Tantayanon

Author Index

Preface

The biennial Pacific Polymer Conference is the official conference co-hosted by the Pacific Polymer Federation (PPF) and the polymer organization of the host country. The PPF was founded in 1987 to encourage and facilitate interactions among polymer organizations of the country members along the Pacific. Current country members are Australia, Canada, Chile, China, Hong Kong, Indonesia, Japan, Korea, Malaysia, Mexico, New Zealand, Singapore, Taiwan, Thailand, the United States (ACS and APS) and Vietnam. Since the establishment of the PPF, the Pacific Polymer Conference has already taken place on eight occasions. On this occasion, the Pacific Polymer Conference was held in the city of Bangkok from November 24 to 27, 2003. The 8th Pacific Polymer Conference (PPC-8) was a tremendous success both in terms of the number of scientific contributions of around 440 contributions including both oral and poster presentations and the number of international participants of around 350 from 31 countries. The scientific programs of the conference consisted of one symposium on hydrogels and nine sessions on functional and synthetic polymers, natural and green polymers, polymer blends and composites, polymer colloids and interfaces, polymer engineering, processing, and characterization, polymer crystallization and structural development during processing, elastomers and rubbers, macromolecular architecture, and industrial and imaging polymers.

All of the selected contributions that are present in this special volume of Macromolecular Symposia are good representatives for manifesting the quality and scientific merit of all of the contributions to the PPC-8. On behalf of the organizing committee, we hope that readers find the information published herein is of use to some extent.

P. Supaphol

S. Tantayanon

Macromol. Symp. **2004**, *216*, 1-8

In-situ Polymerization of Rare-Earth Luminous PA6

Yumei Zhang,* Yunting Ye, Huaping Wang

State Key Laboratory for Chemical Fibers and Polymer Materials, Donghua University, Shanghai, 200051, P. R. China
E-mail: zhangym70@sohu.com

Summary: Luminous polycaprolactam (PA6), synthesized through in-situ polymerization, is a composite of PA6 and uniformly dispersed luminous pigment. During the polymerization, it is found that as luminous pigment was added, the time of prepolycondensation was longer and the molecular weight increased by end-group analysis. It is also indicated that there was chemical coalescence between luminous pigment and PA6 through the end-group analysis and FTIR. From the results of DSC, it is shown that the luminous pigment had large effect on the crystal structure of the luminous PA6. The crystallinity of luminous PA6 varied with the content of the luminous pigment and the heat-treated conditions. From the crystallization kinetics analysis, it is shown that the crystallization rate of the composites increased as luminous pigment added.

Keywords: crystallization behavior; luminous material; polyamides; ring-opening polymerization; structure

Introduction

Research on luminous materials began from the beginning of the 20th century. They are mainly applied for lightening, safety mark and decoration. The development of luminous materials experienced three stages. At first stage, it is sulphide luminous materials including ZnS, CaS and so on, which can give out many different colors. However, the disadvantages of low luminance, short glow time and poor chemical stability make it be used in very small range. In order to increase the glow time, the sulphide luminous material is often contained radioactive element, which would do harm to people's health and environment. In the 1960's, came the second stage. That is the luminous material doped by rare-earth ions. The glow time, luminance and chemical stability are far better than sulphide luminous materials. The rare-earth doped luminous material is also environment-friendly, which brought it great prospect. In the 1990's, came the stage of the application for rare-earth doped luminous material. More practical luminous materials were developed such as luminous plastics, luminous coat and luminous fiber.[1]

DOI: 10.1002/masy.200451201

Combined with the addvantages of polymer materials, polymer-based luminous material is of interest to increase the application of luminous pigments. During the past research work, two ways to prepare luminous polymer products have been reported. First, luminous pigments were blended with the polymer melts or polymer solutions, which is a common and simple method to prepare polymer composites. Some luminous polymer products have been prepared such as luminous PVA, luminous PMMA and luminous viscose fiber.[2] However, there is such a fatal weakness of this method that it is difficult for the luminous pigments to disperse in the polymers, which results to the lower lumanance and poor mechanical properties of the final products. In our preceding work, we take the measure to improve the dispersibility of luminous pigments in polymer melts by surface treatment.[3] Second, luminous pigments were doped with the monomer before polymerization. The luminous pigments can disperse well in the low-viscosity monomers with the strong stir in the polymerization process, which is helpful to improve the processing and machanical properties. Nevertheless, the special chemical structure of the rare-earth element often leads to high reactivity of the luminous pigments, which has large effect on the polymerization.[4] For instance, when the rare earth doped luminous pigments were added in the process of polyester, it was found that the molecular weight of polyester was much lower than that of the common polyester.

Through our work, we find that it is an effective way to prepare luminous PA6 with satisfied molecular weight by adding luminous pigments ($SrAl_2O_4 \cdot Eu^{2+}, Dy^{3+}$) into the monomer before polymerization. However, the polymerization and the structure of the PA6 were influenced by the luminous pigments. Therefore, the polymerization, the molecular weight, the chemical structure, the crystallization behavior as well as the luminous properties of the PA6 composites were investigated in detail.

Experimental Part

Materials

The luminous pigment, $SrAl_2O_4 \cdot Eu^{2+}, Dy^{3+}$, used in this experiment is a laminated photoluminescence material with the particle size of 20μm supplied by Shanghai Yuelong New Materials Co., Ltd. Caprolactam, ω-aminolauric acid, adipic acid and distilled water were commercially available.

Polymerization

The luminous pigment, $SrAl_2O_4 \cdot Eu^{2+}, Dy^{3+}$, Caprolactam, ω-aminolauric acid, adipic acid and distilled water were mixed in a 500ml three-necked separable flask in N_2 atmosphere under normal presure. The mixture was heated and stirred at 140℃ for 30min, and then heated at 250℃ for more than 5 hours. After polymerization, the products were washed with water at 80℃ to remove monomers and then dried at 90 °C under vaccum.

Measurements

The molecular weight of luminous PA6 was determined by end-group analysis[5], in which both amino end groups (NH_2) and carboxyl end groups(COOH) of PA6 were analyzed through a ZDJ-4A chemical electric displacement titrimeter from Shanghai Precise Instruments, Ltd.

FTIR curves were obtained using Nicolet NEXUS-670 spectrometer with the distinguishing rate of 2 cm^{-1}.

The DSC curves were recorded with a DSC822 instrument from Mettler Toledo Company. The samples were heated at the rate of 10℃/min from room temperature to 250℃, and then cooled and heated again at the same rate.[6,7]

The DLI (depolarized light intensity) data were obtained from a GIY-III instrument to analysis the crystallization kinetics of luminous PA6. The samples were melted at 250℃ and then tested at the temperature from 100~180℃.

Result and Discussion

Polymerization

In the bulk polymerization of PA6, the monomers were slowly heated from 140℃ to 250℃ and one hour later it is found that the melts became viscous. However, this phenomenon can not be found until it was kept the temperature at 250℃ for two hours or so when luminous pigment was added. And the contents of luminous pigments larger, the time longer. The relationship of luminous pigment content and the molecular weight is shown in Table 1. An increasing trend can be found as the content of luminous pigment increased.

Table 1. Molecular Weight of PA6 and luminous PA6 from end-group analysis.

Samples	Content of $SrAl_2O_4 \cdot Eu^{2+}, Dy^{3+}$	C_{NH2}	C_{COOH}	Mn
	%	10^{-5} mol/g	10^{-5} mol/g	
PA6	0	5.010	27.62	7642.3
PA6/G-5	5	4.139	21.00	9253.20
PA6/G-10	10	4.537	19.46	11870.2
PA6/G-20	20	4.922	17.78	12995.4

From Table1, it is also found that the end group COOH in luminous PA6 decreased relatively compared with NH_2. It can be deduced that some end groups COOH had chemical interaction with the luminous pigments. It can be verified further from the FTIR spectra. In Figure1(c), it displays stronger peaks in 3400 cm^{-1}, 2940 cm^{-1}, 2880 cm^{-1}, 1650cm^{-1} and 1570 cm^{-1} with the luminous pigment added. To learn whether there is interaction between PA6 and luminous pigments, luminous PA6 were solved in methyl acid and then extracted. From the Figure1(d) of extracted luminous pigment, peaks like –NH- (1570cm^{-1}), C=O(1650cm^{-1}), -CH-(2880 cm^{-1}) and -CH$_2$- (2940 cm^{-1}) can be easily found, while these were not seen from pure luminous pigment in Figure 1(a). It suggests that there is chemical coalescence between luminous pigment and PA6.

Crystal structure

Polyamide is a semi-crystal polymer with various crystal forms at different conditions. For PA6, α crystal is the typical and stable crystal form. In addition, γ crystal appears under the special conditions and it is not so stable as α crystal. At the same polymerization conditions, the luminous pigments have large effect on the crystal structure and crystallinity of PA6. In Figure2(a), two melting peaks can be found in luminous PA6. It suggested the appearance of γ crystal (at lower temperature) besides α crystal (at higher temperature) with the luminous pigments added in PA6. It indicates that luminous pigment has heterogeneous nucleation to PA6 in the polymerization process. However, in cooling and reheating process, these phenomena can not be found. The γ crystal disappeared due to the slowly cooling and reheating.

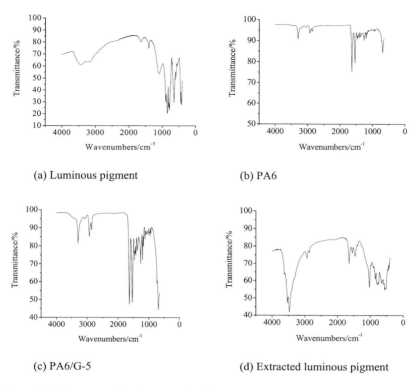

(a) Luminous pigment

(b) PA6

(c) PA6/G-5

(d) Extracted luminous pigment

Figure 1. FTIR spectra of luminous PA6 and luminous pigments.

The crystallinity of luminous PA6 was calculated according to the heat flow from DSC. From Table 2, it is show that the crystallinity of was larger than that of pure PA6 when the samples were heated at the rate of 10℃/min. However, when the heated samples were cooled and heated again, the crystallinity decreased with increasing the content of luminous pigments. The variation of the crystallinity is also caused by the change of crystal form. Actually, there was much more γ crystal in the luminous PA6, which leaded to the increase of crystallinity in the 1st heating process. Meanwhile, the decrease of crystallinity resulted from the disappearance of γ crystal in the cooling and 2nd the heating process.

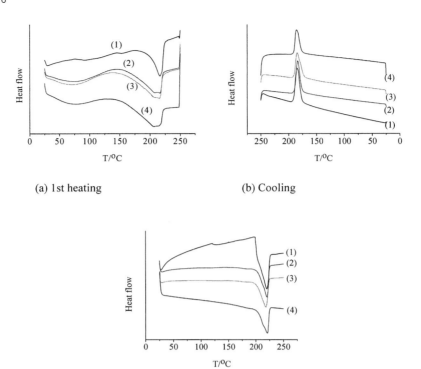

(a) 1st heating

(b) Cooling

(c) 2nd heating

Figure 2. DSC curves of luminou PA6: (1) PA6; (2) PA6/G-5; (3) PA6/G-10; (4) PA6/G-20.

Table 2. Cystallinity of luminous PA6 from DSC.

Samples	Xc/%		
	1st heating	Cooling	2nd heating
PA6	33.84	32.50	37.33
PA6/G-5	40.47	27.85	21.29
PA6/G-10	46.23	24.43	19.06
PA6/G-20	39.48	22.90	16.67

Crystallization Kinetics

The crystallization kinetics of luminous PA6 were studied through DLI (depolarized light intensity). It is found in Figure 3 that the crystallinity rate ($1/t_{1/2}$) decreased as the content of luminous pigments was lower while the crystallinity rate ($1/t_{1/2}$) increased as the content of luminous pigments was higher than 10%. The luminous pigments have special reactivity and special light properties, so the effect of luminous pigments on the DLI data is complex and we will discuss it in other papers.

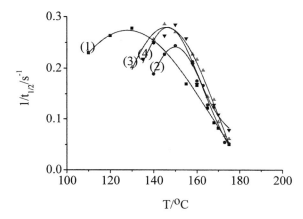

Figure 3. Crystallization Kinetics curves of luminous PA6: (1) PA6; (2) PA6/G-5; (3) PA6/G-10; (4) PA6/G-20.

Luminance

From Figure 4, it is obvious that luminance become largerer as increasing the content of luminous pigment. Especially, the glow time is longer than 1 hour and the luminance is larger than 1.0 mcd/m^2. Therefore, the luminous PA6 is a satisfied material for application.

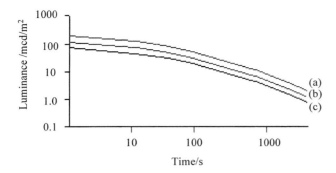

Figure 4. Luminance of luminous PA6.

Conclusion

It is an effective way to prepare luminous PA6 by in-situ polymerization, while the pre-condensation time of luminous PA6 became longer and the molecular weight higher. From the end-group analysis and FTIR spectra, it is duced that there is chemical coalescence between luminous pigment and PA6, which is helpful to the disperse of luminous pigments in PA6 matrix. Under the same conditions, it is easy to form γ crystal in luminous PA6, while it is not found in pure PA6. Luminous pigments play the role of nucleating agent in the crystallization of PA6. Besides, crystallinity of luminous PA6 increased as heated, and decreased while cooling and reheating from DSC. The synthesized luminous PA6 is a satisfied material for application and with the increase of luminous pigments, the luminance of luminous PA6 increased.

Acknowledgements

This research work was supported by Sino Petrochemical Company.

[1]   C. Li , J. Su, J. Qiu, *Chinese Journal of Luninescence* **2003**, *24*, 19.
[2]   X. Ma, *Chinese Journal of Rare Metals* **2000**, *24*, 305.
[3]   J. Wang, Y. Zhang, H. Wang, *Chinese Material Review* **2003**, *6*, 28.
[4]   M. Tang, *OME Information* **2003**, *2*, 8.
[5]   A. Usuki, Y. Kojima, *J. Mater. Res.* **1993**, *8*, 1179.
[6]   Y. Kojima, A. Usuki, *Journal of Polymer Science, Part A, Polymer Chemical* **1993**, *31*, 1755.
[7]   Q.Li, Z. Zhao, Y. Ou, Z. Qi, F.Wang, *Gao Fenzi Xuebao* **1997**, *2*, 188.

Macromol. Symp. **2004**, *216*, 9-16

Free Radical Polymerization of Acrylonitrile in Green Ionic Liquids

Liang Cheng, Yumei Zhang, Tingting Zhao, Huaping Wang*

State Key Laboratory for Chemical Fibers and Polymer Materials, Donghua University, Shanghai, 200051, P. R. China
E-mail: wanghp@dhu.edu.cn

Summary: Free radical polymerization of acrylonitrile (AN) in ionic liquid, 1-butyl-3-methylimidazolium tetrafluoroborate ([bmim][BF$_4$]), 2,2'-azobisisobutyronitrile (AIBN) as initiator was investigated. Early investigations on polymerizations using ionic liquids indicate that they serve as especially good solvents to achieve high molecular weight polymers. Free radical polymerizations result in higher molecular weight polymers, for ionic liquids have low chain transfer constants and act to stabilize the active radical during the process of polymerization. The thermal stability of polymers synthesized in ionic liquids have be improved obviously than that in traditional solvents.

Keywords: acrylonitrile; ionic liquids; radical polymerization; structure; thermal properties

Introduction

Ionic liquids are salts that are liquids at room temperature. Usually composing of a bulky organic cation such as imidazolium or pyridinium cation and a smaller inorganic ion such as PF$_6^-$ or BF$_4^-$, they can be customized like organic solvents. However, ionic liquids are consider "green" because, unlike the volatile organic compounds(VOCs) they reply many of these compounds have negligible vapor pressure and can be recycled and reused repeatedly. Moreover, these liquids have interesting properties that may aid industrial synthesis, such as liquid ranges on the order of 300 ℃ and good thermal stability.

Most of Free radical polymerizations are typically operated in volatile organic solvents, which have been blamed for the increasing pollution. The use of room temperature ionic liquids, particularly 1-butyl-3-methylimidazolium hexafluorophosphate [bmim][PF$_6$] and 1-butyl-3-methylimidazolium tetrafluoroborate [bmim][BF$_4$], as solvents for clean polymerization and

DOI: 10.1002/masy.200451202

catalytic processes, particularly those applicable to clean technology, is becoming widely recognized and accepted[1-4]. Many considerable achievements have been received in the field of free-radical polymerzation, including traditional free-radical polymerization,[5,6] nitroxide-mediated stable free-radical polymerization,[7-10] atom transfer radical polymerization (ATRP),[11-16] reverse ATRP,[17-19] and reversible addition-fragmentation chain transfer.[20-23] The excellent room temperature ionic liquid, [bmim][PF$_6$] as the solvent for the living radical polymerization of methyl methacrylate and acrylates in the presence of a copper(I)/amine catalytic system,[24,25] has been reported.

In this research, we study free-radical polymerization of AN in room temperature ionic liquids, which promises the feasibility that ionic liquids could be used as a green solvent in free radical polymerization. A series of measurements were adopted to analysis effects of ionic liquids and reaction condition on molecule weight, molecular weight distributions and thermal stability.

Experimental Section

Materials

AN (99%) was purchased from Acros Organics and distilled to remove inhibitors by reducing pressure, AIBN (99%) was recrystallized from methanol and dried under vacuum. [bmim][BF$_4$] were synthesized and purified according to the literature in our lab.[26] Dimethyl formamide (DMF) and methanol were purchased from Acros Organics and used without further purification.

Polymerization

The typical experimental procedure is described subsequently. All polymerizations were carried out under nitrogen gas in order to ensure that systems were oxygen free. The desired amounts of initator (AIBN) and solvent ([bmim][BF$_4$]) were added into a round-bottom flask with a magnetic stir bar. Then a predetermined quantity of monomer(AN) was poured into the reactor, the system was purged with nitrogen gas for one hour to displace any surplus oxygen. The flask was placed into a preheated water bath and the polymerization was carried out for 1 hour.

PAN samples were dissolved in DMF solvent and formed 2% solution, which were placed over 24hours. As ionic liquids could not dissolve in DMF, the solution divided to two layers and upper layer were poured away. The left solution was taken shapes of film, then put into pure water about 24hours. Membranes were taken out and dried.

Instrumentation

Gel permeation chromatography(GPC) was used to measure molecular weights and molecular weight distributions, M_w/M_n of PAN samples with respect to polyacrylnitrile (PAN) standards . The GPC experiments with flow rate 1ml/min were carried out at 50℃ in DMF using Series 200 model. Sample concentrations were 1% reaction solution in 99% DMF and measured with a refractive index detector.

Solid-state NMR spectrum of polymers by CP/MAS were recorded on a Bruker ARX-400 spectrometer at room temperature. Fourier transform infrared spectroscopic (FT-IR) analysis were performed on NEXUS-670 model(American Nicolet) FT-IR spectrometer with KBr discs.

Thermal stability and glass transition temperature(Tg) measurements were operated by DSC and TGA. DSC machine was Mettler-Toledo model DSC822 and heating rate is 10℃/min and 20℃ /min. TG2050 machine from American TA was used to determine the rate of losing weight under nitrogen gas with heating rate 10℃/min, the range of temperature is from room temperature to 500℃.

Results and Discussion

Polymers Purification

PAN samples were formed in both DMF and ionic liquids. Formed polymers had the same physical appearance regardless of the solvent in which they were synthesized. The structure of PAN formed in [bmim][BF$_4$] was analyzed with infrared spectrum instrument that is the same as sample formed in DMF. Figure 1 shows that there is an obvious absorption peak at 2242cm^{-1}, which is the character absorption peak of -C≡N. This indicates that the structure of PAN in formed in [bmim][BF$_4$] has combined with the common PAN. However, there exists an obvious difference between the samples A and B. The sample A containing ionic liquids have shorter absorption peak than that B. This could be due to two reasons. Firstly, the polarity of ionic liquids prevents stretching vibration of molecular bonds. As stated earlier, the ionic liquids are highly polar liquids [27,28], which affects the vibration of bonds. Alternatively, existence of ionic liquids in polymers makes better stereo-tacticity than that without ionic liquids, the DSC and TG analysis show the same result.

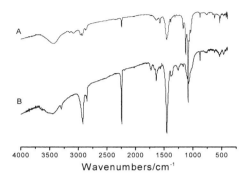

Figure 1. FT-IR spectrum of PAN polymerized in ionic liquids; reaction conditions: T=60 ℃, polymerization time is 60 minutes, 1 wt% initiator (AIBN), 25wt% AN; The sample A includes ionic liquids and that B does not have ionic liquids.

So polymers purification is crucial to product's viability of any solution polymerization process, for a polymer's thermal and mechanical properties are affected by other elements. As reported earlier, solvents traditionally used are volatile and promised to remove by heating or under vacuum, which is easier than ionic liquids because ionic liquids are nonvolatile. However, the existence of ionic liquids could improve some special properties of polymers.

Molecular Weight and Molecular Weight Distributions

The molecular weights of polymer samples were determined as a function of time for each solvent used. The GPC showed that the number-average molecular weight of the polymer was slightly higher than conventional PAN, but the molecular weight distribution is low from Table 1. Polymerization of PAN is a fast reaction when compared to polymerization in popular solvents. The increase in rate is manifested by a widen of the molecular weight distributions to 3.76. Two main reasons could attribute to this result.[29,30] Firstly, the rate of bimolecular termination by either disproportionation or combination is suppressed which might be due in part to the large increase in viscosity of the action medium. In this case termination would be dominated by chain transfer to either solvent or monomer. The molecular weights observed are consistent with this explanation. Alternatively, the rate constant of propagation, k_P, may be increased in the ionic liquid due to local environment effects. Tadeusz[31] and Carmichael[32] reported the similar result and they also considered that either instantaneous concentration of radicals or the propagation-rate constant is higher than in organic media.

Table 1. Final molecular mass and conversion data for PAN synthesized in this work.

Reaction	Temperature/°C	Time /min	Conversion /%	Mn	PDI
1	50	60	90.5	46450	3.65
2	55	60	91.5	44270	3.76
3	60	60	92.2	40760	2.55
4	65	60	93.4	34450	2.32
5	70	60	94.3	32600	3.72

The microstructures of polymer samples were determined by CP/MAS measurements. CP/MAS spectra clearly demonstrates that the polymers were provided with characteristic chemical shifts δ, which could be attributed respectively to $-C \equiv N$ (δ =169.914ppm), -CH- (δ =121.510) and $-CH_2-$ (δ=29.497). Fortunately, it was reported by May[29] that the stereochemistry of the PMMA produced in [bmim][BF$_4$] is almost identical to that for PMMA produced by free radical polymerization in conventional solvents.

Thermal Stability Measurements

Thermal stability is one of important characters for process. The glass transition is important in design of polymeric materials because it marks the transition between hard, glassy properties at lower temperature and rubbery behavior at higher temperatures. It shows by Figure 2 and 3 the polymers with ionic liquids have better thermal stability than those without ionic liquids. Figure 2 shows the T_g of PAN 1 contained ionic liquid dropped significantly, with little change without ionic liquid. Further, the rate of sample 1 losing weight is lower than others without ionic liquid as shown in table 1, which demonstrates ionic liquid could strength thermal stability of polymers. Generally, molecule chains of polar polymers form many physical crosslinked points since polar groups or hydrogen bonds interact strongly. When ionic liquid enter molecular chains, the polar groups of ionic liquid produce effect with that of molecules each other, so which destroy polymers' crosslinked points achieve high sub-chain activity. The procedure could be simulated by following model.

Table 2. The data of rate of PAN TGA.

Sample	1#	2#	3#	4#
Rate of losing weight	22%	28%	33%	43%

1#: PAN with ionic liquids polymerized in [bmim]BF$_4$, monomer concentration 25%(w/w).
2#: PAN polymerized in [bmim]BF$_4$, monomer concentration 25%(w/w).
3#: PAN polymerized in [bmim]BF$_4$, monomer conce ntration 40%(w /w).
4#: PAN polymerized in DMF monomer concentration 25%(w/w).

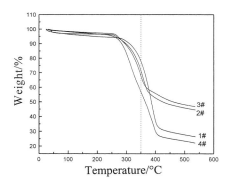

Figure 2. TGA curves of polymers; reaction condition was the same as demonstrated in Table 1.

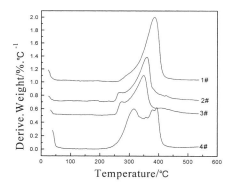

Figure 3. DSC resualts of polymers; reaction condition was the same as demonstrated in Table 1.

Besides, Mark[33] and Cadogan[34] considered that ionic liquids containing imidazolium cations are similar to some conventional plasticizers, containing an aromatic core and pendant alkyl groups. Ionic liquids perform many characteristics similar to plasticizers, such as reproducibility, and wide thermal range being some of the greatest advantage over conventional plasticizing agents. One of the most important characteristics of plasticizer compounds is low volatility so that the polymer can remain flexible over a long lifetime without becoming brittle and failing. Ionic liquids are particularly intriguing in this aspect, as they typically have no detectable vapor pressure, which could enable the formation of flexible materials with significantly extended lifetimes. These results indicate that ionic liquids as plasticizers could strengthen thermal stability of polymers.

Conclusion

PAN was synthesized in [bmim][BF$_4$] using AIBN as initiator, a green alternative to traditional volatile solvents. The effects of reaction parameters on molecular weight and molecular weight distributions of the polymer formed in room temperature ionic liquid are the same as for that observ in traditional solvents.

The polymerization products were characterized by GPC, NMR, FT-IR, DSC, and TGA. The FT-IR analysis shows that the structure of production is the same with the common PAN, however polarity of ionic liquid confines the stretching vibration of molecular bonds. Thermal stability of PAN with ionic liquids is improved obviously because ionic liquids break polymers' crosslinked points, also ionic liquids containing imidazolium cations are similar to some conventional plasticizers.

This investigation of polymerization in ionic liquids sheds light on the potential to develop environmentally-friendly processes and gain an understanding of the reaction behavior to make materials with reduced emission that also have technically superior properties.

[1]   R. T Carlin, *J. S.* Wilkes, *J Mol Catal.* **1990**, *63*, 125.
[2]   V. M. Kobryanskii, S. A. Arnautov, *Chem Commun.* **1992**, 727.
[3]   D. Adam, *Nature.* **2000**, *407*,938.
[4]   T. Welton, *Chem Rev.* **1999**, *99*, 2071.
[5]   M. G. Benton, C. S. Brazel.; *Polymer Preprints.* **2002**, *43*, 881.
[6]   H. Zhang, K. Hong, J. W. Mays.; *Macromolecules.* **2002**, *35*, 5738.
[7]   C. J. Awker, A. W. Bosman, E.Harth, *Chem Rev.* **2001**, *101*, 3661.
[8]   D. Benoit, C. J. Hawker, E. E. Huang, Z. Q. Lin, T. P. Russell, *Macromolecules.* **2000**, *33*, 1505.
[9]   T. Tsoukatos, S. Pispas, N. Hadjichristidis, *J Polym Sci* Part A: *Polym Chem.* **2001**, *39*, 320.

[10]   A. J. Pasquale, T. E. Long, *J Polym Sci* Part A: *Polym Chem*. **2001**, *39*, 216.
[11]   M. Kato, M. Kamigation, M. Sawamoto, T. Higashimura, *Macromolecules*. **1995**, *28*, 1721.
[12]   M. Kamigaito, T. Ando, M. Sawamoto, *Chem Rev*. **2001**, *101*, 3689.
[13]   J. S. Wang, K. Matyjaszewski, *J Am Chem Soc*. **1995**, *117*, 5614.
[14]   K. Matyjaszewski, J. Xia, *Chem Rev*. **2001**, *101*, 2921.
[15]   V.Percec, B. Barboiu, *Macromolecules*. **1995**, *28*, 7970.
[16]   J. S. Wang, K. Matyjaszewski, *Macromolecules*. **1995**, *28*, 7572.
[17]   J. Xia, K. Matyjaszewski, *Macromolecules*. **1997**, *30*,7692.
[18]   G. Moineau, Ph. Dubois, R. Jerome, T. Senninger, Ph. Teyssie, *Macromolecules*. **1998**, *31*, 545.
[19]   S. Zhu, D. Yan, G. Zhang, *J Polym Sci* Part A: Polym Chem **2001**, *39*, 765.
[20]   Y. K. Chong, T. P. T. Le, G. Moad, E. Rizzardo, S. H. Thang, *Macromolecules*. **1999**, *32*, 2071.
[21]   J. G. Tsavalas, F. J. Schork, H. de Brouwer, M. J. Monteiro, *Macromolecules*. **2001**, *34*, 3938.
[22]   M. J. Monteiro, J. de Barbeyrace, *Macromolecules*. **2001**, *34*, 4416.
[23]   Y. Tusjii, M. Ejaz, K. Sato, A. Goto, T. Fukuda, *Macromolecules*. **2001**, *34*, 8872.
[24]   A. J. Carmichael, D. M. Haddleton, S. A. F. Bon, K. R. Seddon, *Chem Commun*. **2000**, 1237.
[25]   T. Biedron, P. Kubisa, *Macromol Rapid Commun* **2001**, *22*, 1237.
[26]   P. A. Z. Suarez, S. Einloft, J. E. L. Dullius, R. F. de Souza, J. Dupont.; *J. Chem. Phys*. **1998**, *95*, 1626.
[27]   S. N. V. K. Aki, J. F. Brennecke, A. Samanta.; *Chem Commun*. **2001**, 413.
[28]   J. G. Huddleston, A. E. Visser, W. M. Reichert, H. D. Willauer, G. A. Broker, R. D. Rogers.; *Green Chemistry*. **2001**, *3*, 156.
[29]   H. Zhang, L. Bu, M. Li, K. Hong, J. W. Mays, R. D. Rogers, *ACS Symposium series chapter, in press* **2001**.
[30]   A. J. Carmichael, D. A. Leigh, D. M. Haddleton, *ACS Symposium series chapter, in press* **2002**.
[31]   T. Biedron, P. Y. Kbisa, *Journal of Polymer Science*: Part A. **2002**, *40*, 2799.
[32]   A. J. Carmichael, D. M. Haddleton, S. A. F. Bon, K. R. Seddon, *Chem Commun*. **2000**, 1237.
[33]   M. P. Scott, C. S. Brazel, M. G, Chem. Commun. **2002**, 1370.
[34]   D. F. Cadogan, C. J. Howick, '*plasticizers*' in *Kirk-Othmer Encyclopedia of Chemical Technology*, ed. J. I. Kronschwitz, M. Howe-Grant. Wiley, New York, **1992**. 258.

Chitosan-Pectin Composite Gel Spheres: Effect of Some Formulation Variables on Drug Release

Pornsak Sriamornsak,[1] *Satit Puttipipatkhachorn*[2]

[1] Department of Pharmaceutical Technology, Faculty of Pharmacy, Silpakorn University, Nakhon Pathom 73000 Thailand
E-mail: pornsak@email.pharm.su.ac.th
[2] Department of Manufacturing Pharmacy, Faculty of Pharmacy, Mahidol University, Bangkok 10400 Thailand

Summary: Chitosan-pectin composite gel spheres were prepared by ionotropic gelation method. Pectin solution containing indomethacin, a model drug, was extruded into a mixture of chitosan and calcium chloride. The release behavior of indomethacin from composite gel spheres was investigated *in-vitro*. The influence of factors affecting release behavior, such as type of pectin, molecular weight of chitosan, cross-linking time and release medium, were discussed in this study. Adding chitosan into gelation medium could retard the release of indomethacin from gel spheres. The different type of pectin used demonstrated slightly different drug release profiles. The higher molecular weight of chitosan showed less indomethacin release than the lower one. The increased cross-linking time slowed the drug release from composite gel spheres. The release of indomethacin from composite gel spheres was also dependent on the release medium. The drug release was slower in tris buffer where no phosphate ions which can induce the precipitation of calcium phosphate. The results suggested that the composite gel spheres of pectin and chitosan could be used as a controlled release drug delivery carrier.

Keywords: chitosan; composite; drug delivery system; gel; pectin; polysaccharides

Introduction

Chitosan is a cationic polysaccharide made from alkaline N-deacetylation of chitin, consists of N-acetylglucosamine and glucosamine residues (Figure 1a). It has biocompatible, biodegradable, nontoxic and mucoadhesive characteristics.[1] Pectin is an anionic polysaccharide that can be used as a carrier for drug delivery to gastrointestinal tract.[2-3] The polymeric chain of pectin contains galacturonic acid and its methyl esters (Figure 1b).

The properties of interpolymeric complex gel beads and films composed of chitosan and pectin have been reported recently.[4-5] The complex formation changed the drug release behavior of hydrogel beads and modified their swelling behavior of complex particles. However, the effect of factor influencing the release behavior, i.e., type of pectin, molecular weight of chitosan, cross-linking time and release medium, has not been examined in detail. The objective of this study was to investigate the release of indomethacin, a model drug, from composite gel particles made of pectin and chitosan.

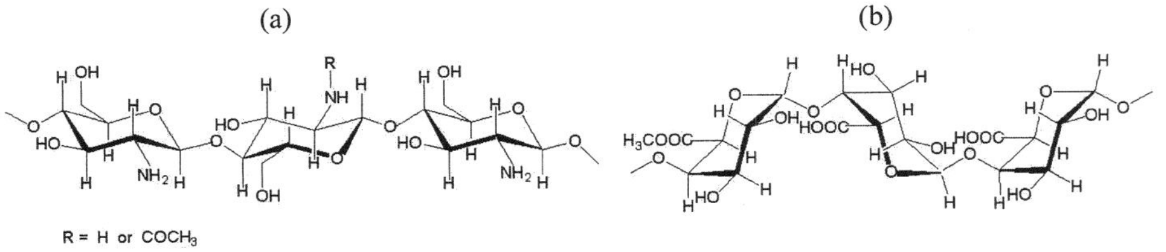

(a)   (b)

R = H or COCH₃

Figure 1. Structure of (a) chitosan and (b) pectin.

Experimental Methods

High (MW 814000) and low (MW 111000) molecular weight chitosans (with degree of N-acetylation of 88% and 95%, respectively) were purchased from Seafresh Chitosan (Lab) Company (Thailand). GENUpectin type LM-101 (degree of esterification, DE 36%) and LM-104 AS-FS (DE 28%) were the generous gift of CP Kelco (Denmark) and are referred to as P36 and P28 respectively. Indomethacin and calcium chloride ($CaCl_2$) were of standard pharmaceutical grade and all chemical reagents used were of analytical grade.

Pectin gel spheres were prepared by ionotropic gelation method.[2-3] Pectin was dissolved (5% w/w) in water with gentle agitation and indomethacin was dispersed to aqueous solution. The dispersions were dropped using a nozzle of 0.80 mm inner diameter into a 5% (w/v) $CaCl_2$ with gentle agitation. The spheres formed were allowed to stand in the solution for 30 min or 24 h, separated and washed with distilled water, then filtered and dried at 37 °C for 12 h. Pectin-chitosan composite gel spheres were prepared by the same method except the mixture of chitosan (0.2% w/v of C-H or C-L) and $CaCl_2$ (5% w/v) was used instead of $CaCl_2$ alone.

The scanning electron microscope (Model Maxim 2000S, CamScan Analytical, England) equipped with back-scattered electron detector and x-ray detector (Model Econ-4, EDAX, USA) was used to examine the surface and cross-section of gel spheres. Drug release kinetics from the gel spheres were evaluated using the rotating basket dissolution method (USP dissolution apparatus 1, Erweka, Germany). The baskets were rotated at 100 rpm at 37 °C. The dissolution medium used was pH 7.4 Tris buffer or pH 7.4 phosphate buffer. All dissolution runs were performed in triplicate.

Results and Discussion

Aqueous solution of pectin containing indomethacin was dropped into either $CaCl_2$ solution or chitosan-$CaCl_2$ mixture and gelled spheres were formed instantaneously by ionotropic gelation in which intermolecular cross-links were formed between the divalent calcium ions and the negatively charged carboxyl groups of the pectin molecules. In case of chitosan-$CaCl_2$ mixture was used, the positively charged ammonium groups of chitosan may also interact with carboxyl groups of the pectin molecules.

Back-scattered electron images revealed the more intense calcium signal from calcium pectinate structure, which appeared brighter than drug particles (Figure 2). On the contrary, the image of chitosan-pectin composite gel sphere (Figure 3) did not show difference in color, suggested that the calcium atoms in the pectin chain have been replaced by positively charged atoms of chitosan. The images of composite gel spheres made of different chitosans showed similar results.

(a)   **(b)**   **(c)**

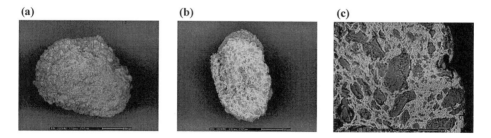

Figure 2. Images of (a) external and (b-c) internal structure of drug-loaded calcium pectinate gel sphere. Magnifications and scale bars are shown on the individual photographs.

(a)   (b)   (c)

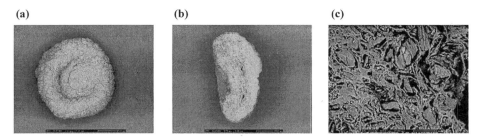

Figure 3. Images of (a) external and (b-c) internal structure of drug-loaded composite gel sphere of chitosan and pectin. Magnifications and scale bars are shown on the individual photographs.

The release behavior of indomethacin from composite gel spheres was investigated *in-vitro*. The influence of factors affecting release behavior, such as type of pectin, molecular weight of chitosan, cross-linking time and release medium, were investigated. Adding chitosan into gelation medium sligthly retarded the release of indomethacin from gel spheres (Figure 4). The different type of pectin used demonstrated different drug release profiles. The lower the DE of pectin showed the slower drug release (Table 1).

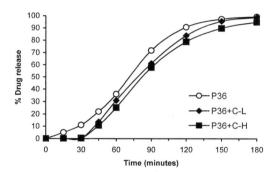

Figure 4. Release (in Tris buffer) of indomethacin from gel spheres. The chitosan used was 0.2% and the cross-linking time was 30 minutes. Note: P36 = calcium pectinate gel spheres using pectin (P36); P36+C-L = composite gel spheres of pectin (P36) and chitosan (low molecular weight); P36+C-H = composite gel spheres of pectin (P36) and chitosan (high molecular weight).

Table 1. Mean percentage of drug release at time of 90 minutes (n=3).

Chitosan	Cross-linking time	pH 7.4 phosphate buffer		pH 7.4 Tris buffer	
		P28	P36	P28	P36
No chitosan	30 min	76.7%	92.9%	25.5%	71.7%
	24 h	80.1%	91.8%	23.2%	45.3%
Chitosan C-L	30 min	60.1%	90.0%	45.5%	61.1%
	24 h	58.2%	80.7%	41.9%	65.6%
Chitosan C-H	30 min	70.6%	90.0%	40.5%	57.6%
	24 h	68.4%	77.5%	32.8%	54.5%

The higher molecular weight of chitosan showed less indomethacin release than the lower one. The increased cross-linking time slowed the drug release from composite gel spheres. The release of indomethacin from composite gel spheres was also dependent on the release medium. The drug release was slower in Tris buffer where no phosphate ions which can induce the precipitation of calcium phosphate. The results suggested that the chitosan-pectin composite gel spheres could be used as controlled release drug delivery carrier.

Acknowledgements

The authors gratefully acknowledge Mr Witoon Sae-Ngow from SUSTREC (Silpakorn University Scientific and Technological Research Equipment Centre) for technical assistance on SEM imaging. Thanks also go to Food & Cosmetic System Co., Ltd. (Thailand) for supplying pectin samples.

[1] H. L. Lueben, C. O. Rentel, A. F. Kotze, C. M. Lehr, A. G. de Boer, J. C. Verhoef, H. E. Junginger, *J. Control. Rel.* **1997**, *45*, 15-23.
[2] P. Sriamornsak, J. Nunthanid, *J. Microencapsul.* **1999**, *16*, 303-313.
[3] P. Sriamornsak, *Eur. J. Pharm. Sci.* **1999**, *8*, 221-227.
[4] O. Munjeri, J. H. Collett, J. T. Fell, *J. Control. Rel.* **1997**, *46*, 273-278.
[5] K. D. Yao, J. Liu, G. X. Cheng, X. D. Lu, H. L. Tu, H. D. Silva, *Appl. Polym. Sci.* **1996**, *60*, 279-283.

Macromol. Symp. **2004**, *216*, 23-35

Polymer Architectures *via* Reversible Addition Fragmentation Chain Transfer (RAFT) Polymerization

Pittaya Takolpuckdee, James Westwood, David M. Lewis, Sébastien Perrier*

Department of Colour and Polymer Chemistry, University of Leeds, Leeds, LS2 9JT, United Kingdom
Fax: +44 113 343 2947; Tel: +44 113 343 2932; E-mail: s.perrier@leeds.ac.uk

Summary: Various versatile chain transfer agents (CTAs) have been synthesized for reversible addition fragmentation chain transfer (RAFT) polymerzation. Such CTAs have been used to modify hydroxyl containing materials and produce well-controlled molecular architectures such as amphiphilic copolymer from poly (ethylene glycol), AB block copolymer consisting of a biodegradable segment, poly (*l*-lactic acid) (PLLA) and grafted copolymers of poly (styrene), poly (methyl methacrylate) and poly (methyl acrylate) from cellulose.

Keywords: block copolymers; cellulose graft polymer; functionalization of polymers; macromolecular architecture; reversible addition fragmentation chain transfer

Introduction

The study of molecular architecture design, resulting in the conception of many novel macromolecular structures such as di-block, grafted, dendrimeric, and hyperbranched polymers, with specific end group functionalities has been widely investigated by living ionic polymerization. However, this technique requires the use of protecting groups and has limited monomer selections.[1] The advantages of radical polymerization over ionic polymerization is the wide range of monomers that can be polymerized and the absence of protecting groups. Moreover, the ability to control polydispersities in a living free radical system makes it a powerful system for polymer synthesis. Living radical polymerization (LRP) techniques such as nitroxide mediated polymerization (NMP)[2], atom transfer radical polymerization (ATRP)[3,4], and reversible addition fragmentation chain transfer (RAFT) polymerization have been extensively studied.[5] RAFT[6,7] and macromolecular architecture design by interchange of xanthates (MADIX)[8,9] (scheme 1) are the newest of the living free radical techniques and one of

          DOI: 10.1002/masy.200451204

their most significant advantages over other techniques is the greater range of polymerizable monomers. The focus of much attention has been the use of both RAFT and MADIX on the key ancillary areas for the synthesis of specific functionalities on polymer backbones with particular chain ends and for the design of various polymeric architectures.

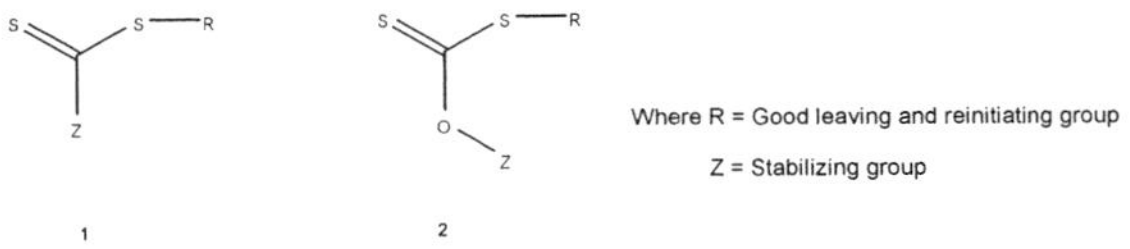

Where R = Good leaving and reinitiating group
Z = Stabilizing group

Scheme 1. Generic formula of RAFT (**1**) and MADIX (**2**) agent.

The use of LRP to synthesize block copolymers with unique properties to be used as compatibilizers or novel specific applications has been widely investigated.[10,11] Of special interest is the synthesis of ATRP initiators from hydroxyl group containing materials reported for the first time by Haddleton et al.[12,13] The study explored a new synthetic route for the preparation of copolymers with specific functionalities and properties which cannot be achieved via normal free radical polymerization methods. More recently, amphiphilic block copolymers of poly (ethylene oxide) with styrene (and other monomers) have been successfully prepared via either ATRP or nitroxide mediated polymerization (NMP) with narrow polydispersities.[11,14-16] In addition, the grafting from cellulose fibers via ATRP has been recently reported by Carlmark et al.[17]

RAFT and MADIX have appeared as the most versatile techniques for living free radical polymerization due to their wide range of polymerizable monomers and ease of scale-up. RAFT and MADIX are based on a similar process, the introduction of a small amount of dithioester of generic formula A, scheme 2 (chain transfer agent, CTA) in a classic free radical system (monomer and initiator). The transfer of the CTA between growing radical chains, present at very low concentration, and dormant polymer chains, present at higher concentration, will regulate the growth of the molecular weight, and limit termination reactions (Scheme 2).

Scheme 2. General accepted mechanism of RAFT.

Here, we report the syntheses of highly versatile CTAs by varying R group to polymeric chains (Scheme 2). The R groups are based on materials that are inexpensive, biodegradable, and renewable resources.

Experimental

Materials

Scoured and bleached, fluorescent brightener-free, woven cotton was used throughout the work. The cotton fabric was dried in a vacuum oven overnight before use. All solvents, monomers and other reagents were purchased from Aldrich Chemical and used as received unless otherwise stated. Tetrahydrofuran (THF), triethylamine and dimethylformamide (DMF) were dried over molecular sieved 4 A. All monomers were purified by passing through aluminium oxide, activated basic Brockmann I. Air and moisture sensitive compounds were manipulated using standard Schlenk techniques under a nitrogen atmosphere.

Characterizations

1. Size exclusion chromatography

Molecular weight distributions were determined using size exclusion chromatography (SEC) at ambient temperature using a system equipped with a guard column and two mixed columns (Polymer Laboratories) with a differential refractive index detector. Tetrahydrofuran was used as an eluent unless otherwise stated at a flow rate of 1 mL min^{-1} with toluene as a flow rate marker. Both poly (styrene) in the range of 7,500,000 and 580 g mol^{-1} and poly (methyl methacrylate) in the range of 1,944,000 – 1,020 g mol^{-1} were used for calibrations.

2. ^{1}H- NMR

^{1}H-NMR spectra were recorded on a Bruker 400 MHz spectrometer with d^3-chloroform as solvent.

3. ATR-FTIR and Raman

IR analysis of cotton was carried out using a Perkin Elmer 1740 Fourier Transform Infrared Spectrometer. The contact sampler used was a horizontal internal reflectance accessory (ATR). One hundred scans were carried out over 222 seconds to produce the final spectrum.

Synthesis and use of Novel Chain Transfer Agents (CTAs)

Preparation of S-Methoxycarbonylphenylmethyl dithiobenzoate (MCPDB) CTA (Scheme 3, CTA 1).

Phenyl magnesium bromide was prepared from bromobenzene (3.14g, 20.0 mmols) and magnesium turning (0.50g, 21.0 mmols) in dry THF. The solution was heated to 40 °C and carbon disulphide (1.525g, 20.0 mmols) was added and a dark brown solution was obtained. Methyl-α-bromophenylacetate (5.00g, 21.8 mmols) was then transferred to the solution. The reaction temperature was raised to 80 °C and maintained for 24 hrs under reflux. Ice water was then added to the solution, before extracting three times the organic products with diethylether. The combined organic extracts were rinsed with saturated sodium hydrogen carbonate and dried over anhydrous magnesium sulphate. After solvent removal, flash column chromatography was undertaken using diethylether:n-hexane (1:9) as eluent.

Preparation of S-methoxycarbonylphenylmethyl methyltrithiocarbonate (MCPMT) CTA (Scheme 3, CTA 2).

The method proposed by CSIRO base on similar Z and R group for the preparation of multi-arm functionalization was used to synthesize MCPMT.[18]

Preparation of S-diethylcarbamoylphenylmethyl dithiobenzoate (DCPDB) CTA (Scheme 3, CTA 3).

α-Chlorophenyl acetyl chloride (8.241g, 0.044 mmols) was added to a round bottom flask containing dry THF and connected to a reflux condenser. 4,4 Dimethyl amino pyridine (DMAP) (0.004 mmols) was then put into the solution. Diethylene amine (0.060 mmols) was then transferred and the solution was allowed to reflux over night. THF was removed and the resultant compound was used without purification.

Phenyl magnesium bromide was prepared as mentioned above in a 3-neck round bottom flask. Carbon disulphide (1.525g, 20.0 mmols) was added dropwise for approximately 15 min and the temperature was increased to 40 °C. 2-Chloro-N,N-diethyl-2phenyl acetamide (4.52 g) that was prepared in the previous step in 20 mL of THF were added to the flask. The orange solid was recrystalized by diethyl ether. The product was dried under vacuum and analyzed with FTIR, Raman and ^{1}H-NMR.[19]

Polymerization sing CTA 1, CTA 2 and CTA 3

The kinetic studies of MCPDB, MCPMT and DCPDB CTAs were processed in bulk systems.[19,20] The ratio of monomer (styrene (Sty), methyl methacrylate (MMA), methyl acrylate (MA) or dimethyl acrylamide (DMA)), CTA and α,α'-azoisobutyronitrile (AIBN) was 5000:10:1 respectively. 1 mL of the solution was transferred in a different ampoule and nitrogen gas was flowed through the solutions for 5 min. The ampoules were placed in a water bath pre-heated to 60°C. Each ampoule was taken out at various times and placed into an ice bath to quench the reaction. The percentage conversions were measured by ^{1}H-NMR. Molecular weights and PDIs were recorded by SEC.

In the case of MCPDB mediated polymerizations, the reactions with each monomer follow pseudo first order rate plot with time and the molecular weight values increase linearly with conversion, following theoretical values, as expected from living polymerization systems.[20] MCPDB gives good control over styrene, MMA, MA, and DMA. Whereas MCPMT and DCPDB also show good control over styrene, MA, and DMA but not for MMA.[19, 20]

Poly(methyl methacrylate)-*block*-Poly(styrene) (PMMA-*b*-PS) was prepared by further reacting a PMMA macro CTA with MCPDB (PMMA-MCPDB) by addition of styrene monomers. The clean conversion of PMMA into block copolymer was confirmed by SEC. The initial PMMA macro CTA (M_n^{SEC} = 13,400 g/mol; PDI = 1.18) was transferred into a PMMA-*b*-PS (M_n^{SEC} = 63,600 g/mol; PDI = 1.27). Whereas the AB block copolymer of PS-*b*-PMA was prepared by the addition of MA monomers into a PS Macro CTA with DCPDB (PS-DCPDB). The chromatograms (Figure 1) show the formation of a PMA second block to form a PS-*b*-PMA (M_n^{SEC} = 42,000 g/mol; PDI = 1.27). This was achieved by addition of MA monomers to the initial PS-DCPDB (M_n^{SEC} = 25,450 g/mol; PDI = 1.23).

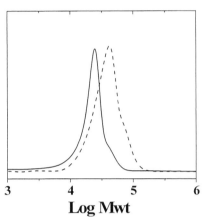

Log Mwt

Figure 1. SEC traces of PS (-) and PS-*b*-PMA (- -) mediated by DCPDB.

Syntheses and Use of Macro CTAs (MCPDB analogues)

As living radical polymerizations mediated by MCPDB[20] have shown a better degree of control over molecular weight than those mediated by MCPMT[20] and DCPDB[19], analogues of MCPDB, scheme 3, were synthesized and subsequently tested in polymerization systems.

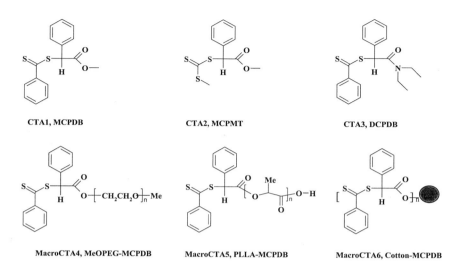

CTA1, MCPDB   CTA2, MCPMT   CTA3, DCPDB

MacroCTA4, MeOPEG-MCPDB   MacroCTA5, PLLA-MCPDB   MacroCTA6, Cotton-MCPDB

Scheme 3. MCPDB, MCPMT, DCPDB and CTA analogues of MCPDB used in this study.

Synthesis of Poly (l-lactic acid) and Poly (ethylene glycol) methyl ester macroCTAs (Scheme 3, CTA 4 and 5)

Poly (*l*-lactic acid) (PLLA)[21] or Poly (ethylene glycol) methyl ester (MeOPEG) (1.81 mmol) was placed in a 3-necked round bottom flask with dry tetrahydrofuran (THF). Triethylamine (2.5 mmols) and 2-chloro-2-phenylacetyl chloride, (CPAC) (2.2 mmols) were added in the flask and the solution was refluxed for 2 days. The solvent was then removed *in vacuo* and dichloromethane was added. The resulting solution was washed with saturated sodium hydrogen carbonate and the yellow organic phase was dried over anhydrous magnesium sulphate. The product was reprecipitated in cold diethyl ether.

The Grignard reaction with carbon disulfide was prepared for the synthesis of MCPDB. However, the 2-chloro-2-phenylacetyl chloride ester of PLLA or MeOPEG synthesised above (21.8 mmols) was then added to the solution instead of methyl-α-bromophenylacetate. The reaction temperature was raised to 80 °C and maintained for 24 hrs. The products were reprecipated in cold diethyl ether.

Preparation of Cotton-MCPDB (Scheme 3, CTA6)

Cotton (4.255 g) was suspended and stirred in a THF solution with triethylamine (72.4 mmols) at 60 °C. CPAC (53 mmols) was added. The reaction was then refluxed for few hours and then cooled down. The light yellow fabric was washed with THF and rinsed thoroughly with deionised water. The treated cotton was oven-dried under vacuum to give the chloro-phenyl-acetic acid cotton ester (cotton-CPA). The degree of substitution (D.S.) was calculated to be 0.53. The cotton CTA was anlyzed by FTIR and Raman. Raman spectroscopy (Figure 2, spectrum A) illustrates the successful reaction of CPAC with the cotton hydroxyl groups by incorporation of the mono-substituted phenyl (1602, 1583 and 1004 cm^{-1}) and the C-Cl bond (617 cm^{-1}) of the acid chloride. Further confirmation was also obtained from FTIR spectroscopy, Figure 3, with stretches at 1736 cm^{-1} and 1187 cm^{-1}, characteristic of C=O and C-O, respectively.

The Grignard reaction with CS$_2$ was prepared as mentioned in the MCPDB preparation. The solution was transferred to another flask containing cotton-CPA (1.923 g). The reaction temperature was then increased to 80 °C and left overnight. The resulting fabric, with a characteristic orange colour, was washed with THF (3 x 20 mL) and then rinsed thoroughly with deionised water. The CTA cotton was then oven-dried under vacuum. D.S. was calculated to be 0.45.

The alteration into dithioester moiety is also described in Figure 3, where the strong Raman bands at 1236 cm^{-1} and 651 cm^{-1} (top spectrum) are accredited to the C=S and C-S bonds, respectively. The characteristic peaks of aromatic structure are still observed at 1602 and 1001 cm^{-1}, demonstrating the incorporation of an extra phenyl ring. The degree of substitution of the hydroxyl groups of the cotton fabric was estimated to be 14% by gravimetric method. The reactions were therefore prepared aiming for a degree of polymerization DP = [M] / [CTA], with [CTA] obtained from the degree of substitution. D.S. is calculated from the number of substituted hydroxyl groups per repeating unit, with a D.S.$_{max}$ = 3.

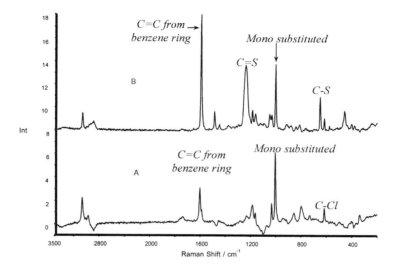

Figure 2. Subtraction of Raman spectra showing [(cotton-CPA) - cotton] (bottom spectrum) and [(cotton-MCPDB) – (cotton-CPA)] (top spectrum).

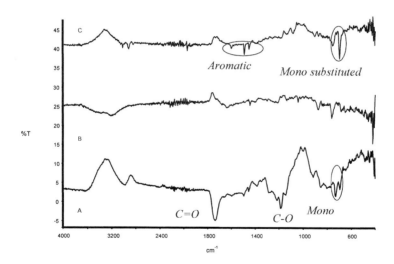

Figure 3. Subtraction of FTIR spectra showing [(cotton-CPA) – cotton].

Polymerizations *via* Macro CTAs

MMA, macro-RAFT agent, and α,α'-azobisisobutyronitrile (AIBN) were added to ampoules in various ratios; 5000:10:1 for PLLA-MCPDB and MeOPEG-MCPDB and 10000:10:1 for cotton MCPDB. Toluene (6 mL) was added to the ampoules. The clinched ampoules were deoxygenated by flushing out with nitrogen for approximately 10 min and placed in a pre-heated oil bath at 60°C. The reactions were stopped at the appropriate times by cooling the reaction tubes in an ice bath. Poly (ethylene glycol) methyl ester-*block*-poly (methyl methacrylate) and poly (*l*-lactic acid)-*block*-poly (methyl methacrylate) were reprecipitated in cold diethyl ether. GPC traces of the initial peaks of PLLA and MeOPEG macro CTAs were shifted after adding MMA, good indication that the second block copolymer is formed.

Table 1. Molecular weight data for AB diblock copolymers and graft copolymers.

Entry	Monomer A	M_n^a	PDI^b	Monomer B	$M_n^{a,c}$	$M_n^{b,c}$	PDI (copolymer)^b
1	Ethylene oxide	5,000	1.12	MMA	43,700	44,150	1.28
2	l-lactic acid	4,800	1.50	MMA	96,300	98,950	1.33
3	Cellulose	-	-	Styrene	-	106,900^d	1.35
4	Cellulose	-	-	MA	-	84,850	1.38
5	Cellulose	-	-	MMA	-	60,500	2.22

^{a} Determined by ^{1}H NMR using CDCl$_3$ as solvent. ^{b} Determined by SEC using THF as an eluent and PMMA standards. ^{c} M_n of the copolymer. ^{d} PS as SEC standards.

^{1}H-NMR spectrum of the final product, shown in Figure 4, displayed the chemical shift signals of both PMMA and MeOPEG macro CTA. The degree of polymerization in this copolymer, estimated from the calculation of methylene protons of PMMA, between 0.5 and 1.3 ppm, with a proton on the phenyl acetyl linkage at 5.24 ppm, correlated well with the GPC results (see in Table 1).

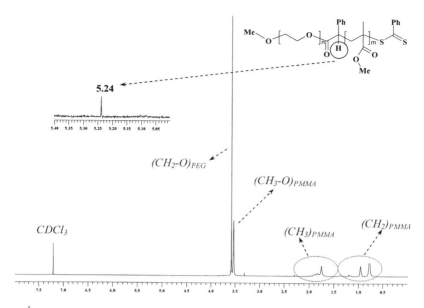

Figure 4. ¹H NMR spectrum for diblock MeOPEG-b-PMMA synthesized from CTA2.

In the cases of cotton polymerizations, after reaction, the grafted cotton substrate was washed 3 times with toluene and twice with THF (while stirring overnight) to remove non-grafted polymeric chains. The grafted cotton sample was hydrolyzed to acidic condition in THF, using 35% hydrochloric acid, to isolate the grafted polymer chains. The solvents were removed in vacuum and water was added to the product to precipitate the polymeric chains. The cotton supported polymerizations of MMA, MA and Styrene were undertaken in bulk up to full conversion. After polymerization, a sample of cotton was hydrolyzed in acidic conditions to yield the grafted PMMA, PS and PMA chains which were then analyzed by SEC. Table 1 exemplifies the remarkable control obtained during polymerization. PS and PMA (entry 3 and 4) exhibit a molecular weight very near to the expected one ($M_{n,\ theo}$ = 104,200 and 99,100 g mol^{-1}, respectively) with polydispersities as low as 1.3. In the case of PMMA (entry 5), a higher molecular weight than anticipated was obtained ($M_{n,\ theo}$ = 100,100 g mol^{-1}), with higher PDI. Therefore, an easy process for supported polymerization was demonstrated, which leads to a remarkable control of the polymeric chains molecular weight. Furthermore, this technique brings new scope in the grafting and functionalization of natural textiles. Additional studies to improve the current technique are currently undertaken in our laboratories.

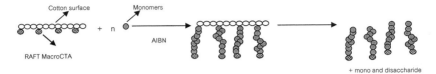

Scheme 4. Hydrolyzed the graft copolymer on the cotton fabric.

Conclusions

We have described the facile synthesis of highly versatile CTAs that can produce an extensive range of polymers with predictable M_n and low PDI and that allow the easy incorporation of any molecules containing a hydroxy group at the end of a polymeric chain. We believe this technique will bring new opportunities in the synthesis of functional polymers, so far dominated by atom transfer radical polymerization. The major advantage of RAFT over ATRP is that potential pollution of the final product by the catalyst is avoided. In addition the process is easier to undertake, with a wider range of monomers.

Acknowledgements

P.T. acknowledges the Royal Thai Government for his financial support. The authors thank Prof J T Guthrie for his advice on the cellulose work.

[1]   K. Matyjaszewski *Curr. Opin. Sol. State Phys.* **1996**, *1*, 769.
[2]   C. J. Hawker, A. W. Bosman, E. Harth *Chem. Rev.* **2001**, *101*, 3661.
[3]   M. Kato, M.Kamigaito, M. Sawamoto, T. Higashimura Macromolecules **1995**, *28*, 1721.
[4]   J. S. Wang, K. Matyjaszewski *J. Am. Chem. Soc.* **1995**, *117*, 5614.
[5]   K. Matyjaszewski *ACS Symp. Ser.* **2000**, 768.
[6]   T. P. Le, G. Moad, E. Rizzardo, S. H. Thang, In PCT Int. Appl. WO 9801478 A1 980115, **1998**.
[7]   J. Chiefari, Y. K. Chong, F. Ercole, J. Krstina, J. Jeffery, T. P. T. Le, R. T. A. Mayadunne, G. F. Meijs, C. L. Moad, G. Moad, E. Rizzardo, S. H. Thang, Macromolecules **1998**, *31*, 5559.
[8]   P. Corpart, D. Charmot, T. Biadatti, S. Zard, D. Michelet, In WO 9858974, 1998.
[9]   D. Charmot, P. Corpart, H. Adam, S. Z. Zard, T. Biadatti, G. Bouhadir, *Macromol. Symp.* **2000**, *150*, 23.
[10]   S. Kawaguchi, M. A. Winnik, K. Ito, *Macromolecules* **1995**, *28*, 1159.
[11]   F. J. Hua, Y. L. Yang, *Polymer* **2001**, *42*, 1361.
[12]   D. M. Haddleton, A. M. Heming, D. Kukulj, S. G. Jackson, *Polymeric Materials Science and Engineering* **1999**, *80*, 81.
[13]   D. M. Haddleton, C. Waterson, *Macromolecules* **1999**, *32*, 8732.

[14]   M. Even, D. M. Haddleton, D. Kukuji, *Europ. Polym. J.* **2003**, *39*, 633.
[15]   B. Y. K. Chong, T. P. T. Le, G. Moad, E. Rizzardo, S. H. Thang, *Macromolecules* **1999**, *32*, 2071.
[16]   H. D. Brouwer, M. A. J. Schellekens, B. Klumperman, M. J. Monteiro, A. L. J. German *Polym. Sci. Pol. Chem.* **2000**, *38*, 3596.
[17]   A. Carlmark, E. E. Malmstroem, *Biomacromolecules* **2003**, *4*, 1740.
[18]   R. T. A. Mayadunne, G. Moad, E. Rizzardo, Tetrahedron Lett. **2002**, *43*, 6811-6814.
[19]   P. Takolpuckdee, S. Perrier, unpublished data.
[20]   S. Perrier, P. Takolpuckdee, J. Westwoods, D. M. Lewis, *Macromolecules*, submitted **2003**.
[21]   S. I. Moon, C. W. Lee, M. Miyamoto, Y. J. Kimura, *Polym. Sci. Pol. Chem.* **2000**, *38*, 1673.

Macromol. Symp. **2004**, *216*, 37-46

Non-Isocyanate-Based Polyurethanes Derived upon the Reaction of Amines with Cyclocarbonate Resins

Constantinos D. Diakoumakos,* Dimiter L. Kotzev

New Technologies, Research & Technology Department, Adhesives & Tooling, HUNTSMAN Advanced Materials, Duxford, Cambridge CB2 4QA, United Kingdom
E-mail: konstantinos_diakoumakos@huntsman.com

Summary: Laprolate-803 (L-803) was used as the model cyclocarbonate resin for reactivity and kinetic studies of non-isocyanate-based solventless polyurethanes derived upon the reaction of amines with the 1,3-dioxolan-2-one (dioxolanone) rings of L-803. It was shown that the reaction could undergo either acid-, base- or metal-type catalysis yielding polyurethanes with 100% conversion of the dioxolanone rings at approximately half the time needed for their uncatalyzed counterparts. The activation energy of the reaction of L-803 with diethylene triamine in the presence or absence of a catalyst was determined via Arhenious studies. The reactivity of amines towards the aforementioned model resin was correlated to their chemical structure. A model non-isocyanate-based polyurethane displayed T_g of -1°C (DMA), fast gel time (390 min) at room temperature and good elongation (~70%).

Keywords: cyclocarbonate resins; mechanical properties; non-isocyanate based polyurethanes; polyurethanes; resins

Introduction

Nowadays, processes and materials that are involved or are suspected to be involved in negative environmental effects are under serious revision and alternative ecologically safer solutions are demanded. Cyclocarbonates are a relatively new class of compounds attracting research interest due to their potential use in the preparation of "green" (the highly toxic isocyanates and their predecessors (phosgene) are not involved in the process), porous-free and moisture insensitive polyurethanes. Cyclocarbonates can be synthesized from corresponding epoxy precursors. Network or linear non-isocyanate-based polyurethanes can be formed from the reaction of cyclocarbonate resins and amines[1-10]. No volatile or non-volatile by-products are produced via this reaction resulting to porous-free polyurethanes and an intramolecular hydrogen bond (see Scheme 2) seems to be responsible for lowering the susceptibility of the backbone to hydrolysis

DOI: 10.1002/masy.200451205

resulting in substantial increase of the chemical resistance. The present work focused on the kinetics, and the effect the amines' chemical structures can have on them.

Experimental

Raw Materials

Laprolate-803 (Scheme 1) (L-803, liquid resin bearing 100% 1,3-dioxolan-2-one groups, MW=927) manufactured by *Macromer Ltd.*, Russia was provided by Chemonol Ltd (Israel), Q19262 (4,7,10-trioxa-1,13-tridecanediamine) and Jeffamine D-230 (JD-230) were provided by VANTICO A.G. (now HUNTSMAN Advanced Materials LLC.). Metatin catalysts 702®, 812® and S-26® were obtained by KMZ Chemicals. DABCO-33LV was provided by Air Products. Ethacure-100® (diethyltoluene diamine) and Ethacure–300® were provided by Albemarle. Diaminodiethyldiphenylmethane (DDDM) was provided by Kayahard AA. The rest of the chemicals used in this work (amines, imines, tetrabutyl ammonium bromide, glacial acetic acid, and methanesulfonic acid) were all purchased from Aldrich.

Analytical Methods and Techniques

Attenuated total reflectance fourier transform infrared (ATR FT-IR) spectra were recorded on a Nicolet 510 FT-IR spectrometer. Differential scanning calorimetry (DSC) was carried out on a DSC-2920 (TA Instruments), in air atmosphere at a heating rate of 10°C/min. Thermal mechanical analysis (TMA) was carried out on a TMA-40 (Mettler) at a heating rate of 10°C/min. Thermal gravimetric analysis was carried out on a Mettler TG50 thermo gravimetric analyser equipped with a Mettler M3 thermo balance. The samples were heated from 50 up to 1000°C (heating rate: 20°C/min) in air atmosphere (IDT: initial decomposition temperature where 2% of weight loss was recorded, PDT_{max}: maximum polymer decomposition temperature, Y_{char} @ 500°C and Y_{char} @ 600°C: char yields at 500 and 600°C). The AR-2000 dynamic analyzer (TA Instruments) equipped with torsional rectangular fittings (specimen: 55 mm in length, 10 mm in width and 2 mm in thickness) was used for dynamic mechanical analyses (DMA) (strain: $\pm 1\%$, frequency 1Hz). Isothermal moisture uptake measurements were carried out at 22°C (acc.: $\pm 2^{\circ}$C) and relative humidity 73% (acc.: $\pm 2\%$) for 60 days. The mixture of cyclocarbonate resin (L-803) and amine (DETA) reacted stoichiometrically in an aluminium dish placed in a desiccator at ambient temperature for 1 day and subsequently they were put in an oven for 2 days at 60°C. Upon that, the dish containing the polyurethane, was placed in a temperature and humidity controlled chamber and weigh measurements were carried out periodically for 60 days. Gel time measurements at 25°C were carried out on a Michler apparatus equipped with a digital temperature controller (acc.: ±0.1°C). Measurements of reaction's exothermic profiles (temperature vs. time) at room temperature were carried out in a Dewar beaker fitted with a digital thermometer equipped with a K-type probe. Tensile and tear strength measurements at ambient temperature, were performed on an Instron 4467 according to ISO R527/2 and DIN 53356, respectively. Shore hardness D measurements were carried out on a model BS660 power assisted operating stand equipped with DuroProbe model BD-2 (Hampden Test Equipment Ltd., U.K.), according to DIN 53505.

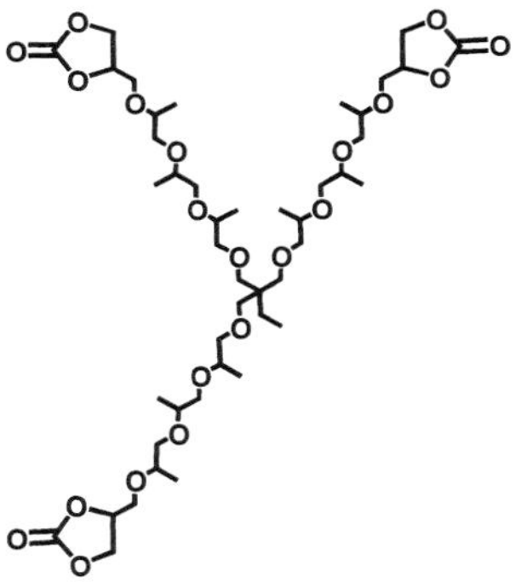

Scheme 1. Chemical structure of L-803.

Results and Discussion

Kinetic Studies and Catalysis

In the present investigation, Laprolate-803 (L-803) was used as a model cyclocarbonate resin for reactivity and kinetic studies of non-isocyanate based polyurethanes derived from the reaction of amines with resins bearing 1,3-dioxolan-2-one rings (cyclocarbonated resins) (Scheme 2).

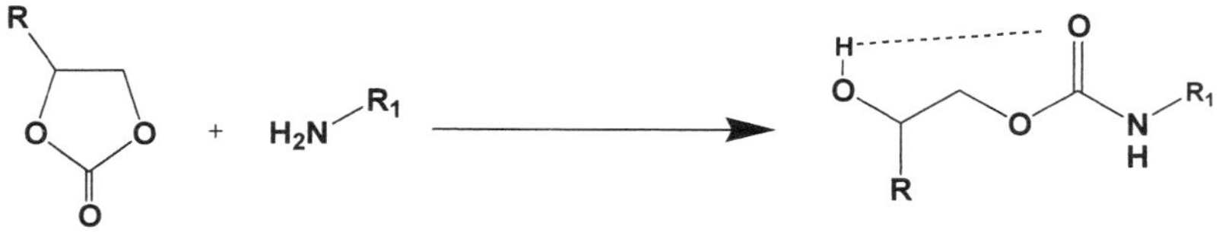

Scheme 2. Reaction scheme between an amino group and a 1,3-dioxolan-2-one ring.

In the reaction of amines and cyclocarbonated resins there is no detailed study addressing in particular the chemical structure of the amine to its reactivity towards the dioxolanone ring. In addition, and although the reaction proceeds satisfactorily at room temperature, there are no in-depth analyses on the reaction's kinetics and potential catalysis for speeding up relatively slow systems as those where L-803 is the only cyclocarbonated resin involved in the reaction. Therefore, a variety of aliphatic and aromatic di- or poly-amines were reacted in bulk -solventless reactions- under the same conditions in the presence or absence of different catalytic systems (e.g. triethylamine, piperazine, tetrabutyl ammonium bromide, glacial acetic acid, methanesulfonic acid, DABCO-33LV and metatin type substances like the Metatin-702[®], -812[®] and -S-26[®]) at 25 and 60°C. As a model polyurethane system for evaluating the activation energy (E_a) of the reaction between the dioxolanone ring and primary aliphatic amino groups via an Arhenious study based on FT-IR spectroscopy, L-803 stoichiometrically reacting with DETA (model system) was selected. L-803 resin shows a very characteristic absorption at 1795 cm^{-1} (C=O, stretching) whereas the urethane carbonyl absorbs (stretching vibration) at 1701 cm^{-1}. The decrease in the intensity of the carbonyl of the dioxolanone ring and at the same time the increase in the intensity of the absorption of the urethane carbonyl were attributed to the formation of polyurethane. The increase of the intensity of the absorption at 1701 cm^{-1} was recorded as a function of time for four different reaction temperatures (25, 60, 80, 100°C) in the presence and absence of triethylamine (catalyst). Figure 1, presents the Arrhenious plots for the reaction of L-803 and DETA in the absence and presence of triethylamine after 30 min. of reaction time. In the case where no catalyst was used the activation energy of the reaction was equal to 6.33 KJ/mol. The low E_a of the reaction permits the initiation of the addition-type polymerization to proceed very fast resulting to an abrupt increase of the viscosity of the mixture inhibiting the completion of the reaction of all the dioxolanone rings with amino groups at relatively low conversion. The reaction was completed after 8 days (no absorption at 1795 cm⁻¹) at ambient temperature, whilst 4 days were necessary for the reaction's completion at 60°C. When triethylamine (1% w/w on total reactants' weight) was used as a potential catalyst the activation energy of the reaction was found to be 5.23 KJ/mol. Although, the activation energy of the uncatalyzed reaction is low, the introduction of triethylamine is beneficiary as it contributes to a further decrease of the activation energy by approx. 17.5% and halved the reaction times necessary for 100% conversion at both 25 and 60°C. The relatively low activation energies determined for the reaction may safely lead us to the assumption that the reaction starts as soon as the two reactants come in contact.

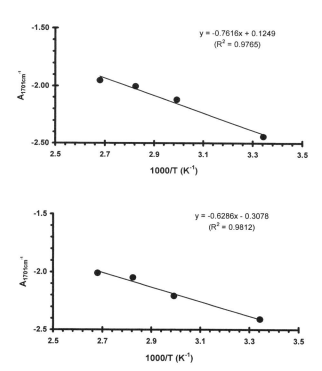

Figure 1. Arrhenious plots for the reaction of L-803 and DETA in the absence (top) and presence. (bottom) of Et₃N as catalyst.

This was also verified with *in-situ* FT-IR studies of a reaction mixture consisting of L-803 and DETA in the absence and presence of triethylamine (Figure 2) as a function of time. The curve shifts towards lower reaction times, when triethylamine was used as catalyst. The same pattern was recorded at 60°C. Different catalysts –as those mentioned here before- were also tested similarly and their effectiveness was assessed via FT-IR spectroscopy based on the decrease of the absorption at 1795 cm^{-1} over a certain time period.

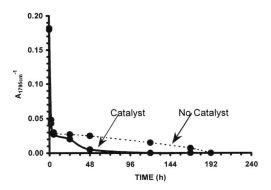

Figure 2. Intensity of the absorption peak at 1795 cm^{-1} as a function of time for the reaction of L-803 and DETA at ambient temperature, in the absence and in the presence of Et$_3$N as catalyst.

It seems that the reaction can undergo various types of catalysis (e.g. acid or base or metal type catalysis) as the array of the different catalysts tested herein gave evidence that they can catalyze it. From all the substances tested as potential catalysts triethylamine, piperazine, tetrabutyl ammonium bromide, glacial acetic acid, methanesulfonic acid and metatin type substances emerge as the most efficient catalysts providing polyurethanes with 100% conversion of the dioxolanone rings to urethane groups after 3-4 days at ambient temperature and 1-2 days at 60°C.

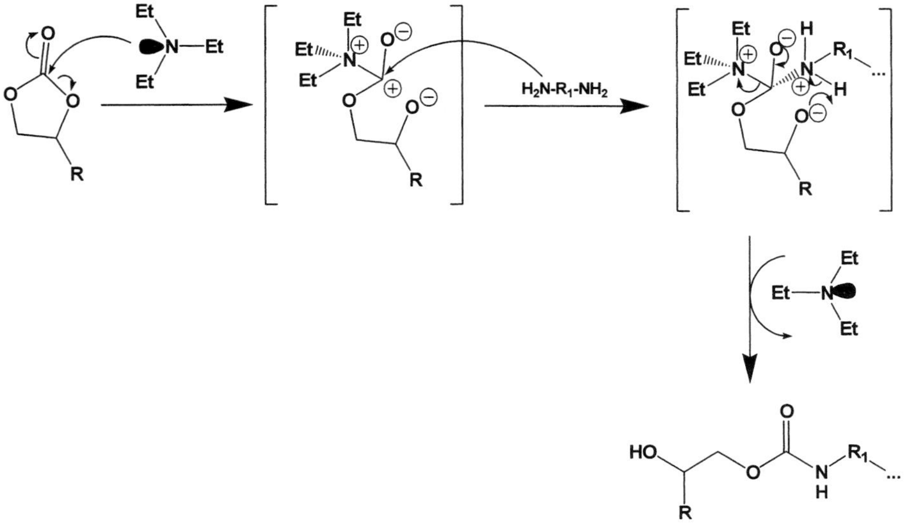

Scheme 3. Proposed catalytic mechanism in the case of triethylamine (base type catalysis).

Reactivity of Amines

A series of experiments were carried out at ambient temperature ($T_{ambient}$= 22-25°C), involving temperature measurements as a function of time for three different reaction masses (11.62, 17.43 and 22.86 g) of our model system (L-803/DETA). The trend of the exothermic profiles recorded in all the reaction mixtures were the same (Figure 3). The presence of triethylamine did not differentiate either the trend of the curve or the maximum temperature in any of the experiments. The maximum increase in temperature $DT=(T_{max}-T_{ambient})$, of these three reaction mixtures was 10, 15 and 19°C, respectively.

Linear curve fitting analysis (T_{max} *vs.* Reaction mass) provided data for calculating the highest temperature that a given mass of the aforementioned reactants can produce.

$$DT=(T_{max}-T_{ambient})= 0.774 \cdot W_{Reactants}+ 1.167 \quad (1)$$

A reaction mass of ca. 170 gr. was the highest reaction mass [much higher than those used to produce equation (1)] that used to verify equation (1) and therefore this is the upper limit of a reaction mass that it was found to comply with the above equation.

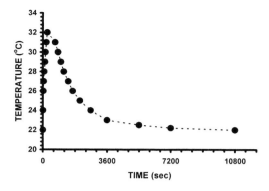

Figure 3. Exothermic profile (temperature vs. time) of a reaction mixture (11.62 g) L-803 and DETA at ambient temperature.

Similar methodology involving reaction masses of the same scale as those used in the case of L-803 and DETA) was also applied in the case of other amines (e.g. TEPA) resulting to equations similar to equation (1). In addition, linear curve fitting analysis taking into account only the mass of the amines (T_{max} vs. Amine's mass) can provide comparative measures (slope and T_{max}) related to their reactivity towards L-803. This methodology in combination to reactivity test experiments (L-803 reacted with various amines) assessed via FT-IR studies (recording the changes in the intensity of the peak at 1795 cm^{-1} vs. time) helped us to correlate an amine's chemical structure to its reactivity towards L-803 (cyclocarbonated resin). It became evident that the reactivity of the amines towards the L-803 is basically governed by two major parameters: a) its chemical structure and b) its molecular weight. Being more precise, the reactivity of the amine is directly associated to: a) the existence of bulky and/or strong electron withdrawing groups in a- or b-position in respect to the reactive amino group and b) the co-existence of imino and amino groups (polyamines) preferably positioned along an aliphatic carbon chain. Scheme 4 depicts in detail all our findings regarding the reactivity of various amines in respect to their chemical structures towards L-803 resin.

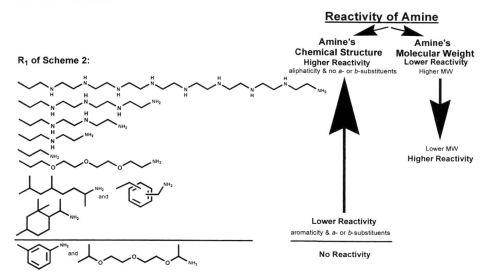

Scheme 4. Reactivity of various amines towards L-803 resin.

Certain typical properties of a model polyurethane (L-803/DETA) derived upon the reaction of L-803 and DETA (model polyurethane) for 4 days at ambient temperature in the presence of 1% w/w Et_3N are mentioned hereafter and in Figure 4: Gel time: 390 min, (DSC): T_g= -20°C, IDT= 272°C, PDT_{max}= 360°C, Y_{char} @ 500°C= 88%, Y_{char} @ 600°C= 11%, (TMA): T_g= -14°C, Coefficient of thermal expansion (CET)= 55.20×10^{-6}/K (below T_g) and 263.00×10^{-6}/K (above T_g), (DMA): T_g= -1°C, Isothermal moisture absorption: 25.90%, (ISO R527/2): Maximum Tensile strength: 0.48 MPa, Young's Modulus= 16.40 MPa, Maximum Load: 11.40 N, Elongation at break= 72.30%, (DIN 53356): Tear strength= 1.75 N/mm, Average Peel Load= 7.20 N, Maximum Load= 8.70 N, Shore hardness D= 11.00.

46

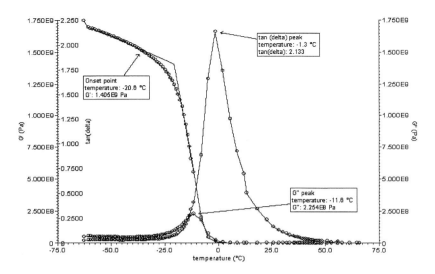

Figure 4. Dynamic mechanical analysis of the model polyurethane.

Conclusion

Cyclocarbonate resins offer a reliable alternative for the preparation of environmentally friendly non-isocyanate based, porous-free and moisture insensitive polyurethanes.

Acknowledgement

The authors wish to thank Professor Oleg Figovski and Dr. Leonid Shapovalov of Chemonol Ltd. (Israel), for providing them with Laprolate-803 (L-803) resin.

[1]   A. E. Gurgiolo, W. L. Bressler, J. C. Smith, **1963**, US. Pat. No. 3,084,140.
[2]   J. W. Whelan, R. J. Cotter, **1963**, US. Pat. No. 3,072,613
[3]   V. V. Fedorova, O. L. Figovsky, I. Kreindlina, E. V. Grigor'ev, **1976**, SU 529197
[4]   V. V. Micheev, N. V. Svetlakov, R. M. Garipov, V. A. Sysoev, R. R. Gil'manov, N. G. Gafiatullin, K. Kirova, *Lakokrasochnye Materialy I Ikh Primenenie* **1985**, 6, 27-30.
[5]   L. Rappoport, R. D. Brown, **1992**, U.S. Pat. No. 5,175,231.
[6]   W. C. Crawford, E. T. Marquis, H. P. Klein, **1994**, U.S. Pat. No. 5,340,889.
[7]   O. L. Figovsky, **1999**, *WO 9965969*
[8]   O. L. Figovsky, L. D. Shapovalov, N. Blank, **2000**, *EP Pat. No. 1020457*
[9]   O. L. Figovsky, L. D. Shapovalov, *Surface Coatings Australia* **2000**, *37*, 14-17
[10]   O. L. Figovsky, L. D. Shapovalov, *Macromolecular Symposia* **2002**, 187.

Macromol. Symp. **2004**, *216*, 47-54

Synthesis and Characterization of β-Poly(glucose-amine)-N-(2,3-dihydroxypropyl) Derivatives as Medical Care and Biological Joint Material. Family 2. Tri or Tetra-Sulfated β-Chitosan

Ryu Ga Youn,[1] *Ryu Soung Ryual,*[*][1] *Jang Hang Doung,*[2] *Jo Byeong Uk*[3]

[1] Department of Chemical Engineering, Dae Bul University, Young-am gun, Chonnam, 526-702, Korea
E-mail: ryusr@mail.daebul.ac.kr
[2] Department of Chemical Engineering, Seoul National university of Technology, Nowaun Gu, Seoul, 139-743, Korea
[3] Department of Chemical Enginerring, Chosun University, Kwang Ju, Chonnam, 501-759, Korea

Summary: Chitin is a natural polysaccharide by N-acetyl-D-glucosamine units β(1-4) linked. In the present work a chitosan with DA 56% and Mv 80.000 g/mol will be employed. Several techniques to obtain sulphate derivatives of chitin and to chitosan have been proposed due to the interesting biological and chemical properties of such as heparin compounds. Among others it could be mentioned their antibactericidal and metal chelating properties. In the present work a new sulfated derivatives of chitosan with potential growth regulator properties were obtained. The reaction was carried out in heterogeneous media using a commercial growth regulator -SO$_3$H as sulfating agent. All derivatives were characterized by several spectroscopic techniques. In IR spectra the bands at 1240 cm^{-1}, (S=O stretching) and 860 cm^{-1}, corresponding to S-O-C symmetric stretching, are characteristic of sulphates groups. In ^{1}H-NMR spectra the appearance of functional group signal confirm its structure.

Keywords: β-chitosan derivatives; sulfated chitosan; tri or tetra-sulfated β-chitosan

Introduction

Chitin [poly(1→4)-2-acetamido-2-deoxy-β-D-glucopyranose][1] is one of the important biomass resources, but the physicochemical and biological characteristics have not been fully disclosed yet owing to the intractable nature. α-chitin has been studied most extensively because of the abundance and easy accessibility. Although only little attention has been paid to *β-chitin*. It may be a promising alternative source of chitin with distinctive features. *β*-Chitin is characterized by weak intermolecular forces[2] and has been confirmed to exhibit higher reactivity in various modification reactions as well as higher affinity for solvent than α-chitin.[3] It is noteworthy that *β*-chitosan prepared from β-chitin also exhibited high reactivity

DOI: 10.1002/masy.200451206

compared to that from α-chitin and significant bactericidal activity.[4-7] These results suggest the high potential of chitosan derived from β-chitin as a novel functional biopolymer.

β-Chitosan is obtained by alkaline N-deacetylation of β-chitin. β-Chitosan are generally insoluble in water because of the strong intermolecular hydrogen bonding. However, they can be converting to water soluble or organic solvent soluble chitosan derivatives by chemical modification on amino, hydroxyl group of glucopyranose ring. Chitosan derivatives are generally recognized as non-toxic biomaterials with good biocompatibility. They have attracted much interest in biomedical application such as wound healing, anti microbial agents, cholesterol reducing agent, and drug delivery carrier, burn healing bio-dressing, drug delay-releasing material and blood vessel disease medicine, and food for lower blood-fat. etc. Here we report the formation of β-chitosan sulfonate derivatives prepared from our laboratory were compared with their physical property, chemical spectroscopy, ets. The degree of deacetylation (DA) and molecular weight (MW) of chitosan were determined by titration of sodium hydroxide solution and by gel permeation chromatography. These modified derivatives included sulfonate. In a current study, sulfated chitosan derivatives were synthesized to utilize chemical properties of chitosan. In addition, chitosan, as a α,β-dihydroxy propionyl[DHP] group substitution, was applied using α,β-diacetatepropionic aldehyde[DAPA]. A [DHP] contained β-chitosan derivative with large surface area -OSO_3H was then produced in order to act as heparin chemical action in solvents.

Experimental

Materials

β-Chitosan was isolate from squid (*Dasidicus gigas*)pens was prepared in our laboratory as described alkali hydrolysis method. Briefly, β-chitin were treated with 40% NaOH at 100 °C for 2 h. filtered, and washed with water. The deacetylation procedure was repeated three more times to give β-chitosan. Other chemicals : acrolein, potassium permanganate, and NN'-dicyclohexylcarbodiimide(DCCI) were purchased from Wako Chemical Co., Japan. Characterization of structural changes in β-chitosan and its derivatives were determined by the Nicolet 5DX FT-IR spectrophotometer and ^1H NMR spectra were recorded on a Varain T60 and HA-100 spectrophotometer.

Preparation of N-(2,3-dihydroxy)propyl Chitosan Derivative (2)

β-Chitosan powder (MW 143K) 1.0 g was dissolved in 50 ml 2% (w/v) acetic acid solution, followed by adding 50 ml MeOH to prepared *dihydroxy)propyl-β*-chitosan solution. 3.0 g of 2,3-diacetylpropionicaldehyde dissolved in 30 ml methanol were added dropwise to the *β*-chitosan solution. After stirring for 12hrs at 25 ℃, the mixed solution was precipitated with 300 ml of MeOH and stirred overnight. Finally, the product was centrifuged, dialyzed with water and dried by lyophillization. and this *β*-chitin derivatives were treated with 40% NaOH at 100 °C for 2 h. filtered the solution was precipitated by adding 500 ml MeOH, followed by centrifugation and was dialyzed with distilled water. The dried product was obtained by $CaCl_2$ contained vacuum oven.

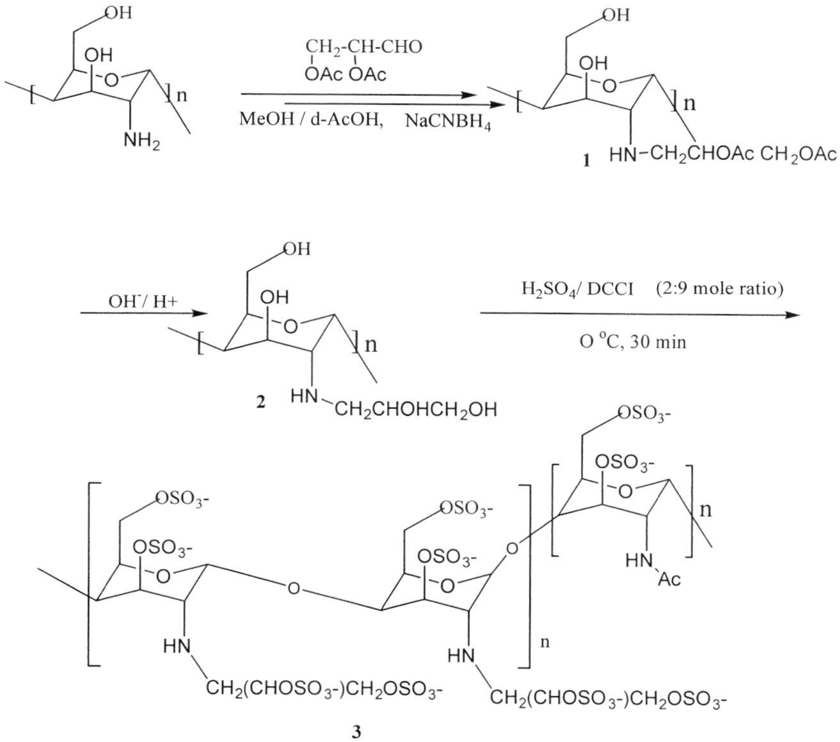

Scheme 1. Reaction pathway for synthesis of chitosan derivatives containing o-sulfate anion.

IR (KBr) spectrum shows λ_{max} (Cm^{-1}) at : 3443-3200(OH), 3240(NH), 2983, 2856(CH), 1450-1420(CH_2), 1368, 1062.

^{1}H-NMR (20% DCl/D_2O): 1.46(d, 2H, -CH_2-), 1.84(t, 1H, =CH-), 2.7(d, 2H, -CH2-), 3.68(d, 2H, -CH_2-), 3.8-4.1 (H2, H3, H4, H5, H6)

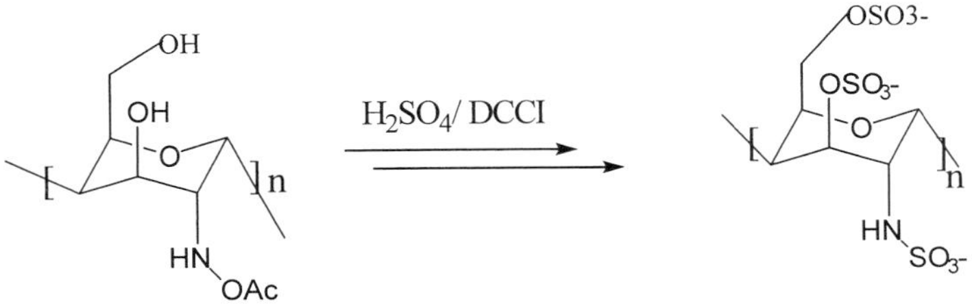

Scheme 2. Reaction pathway for synthesis of chitin derivatives containing o-sulfate anion.

Sulfonation of β-Chitosan Derivatives (SC)

β-Chitosan derivatives(2) (MW 183K) 1.5 g was dissolved in 50 ml of DMF: 2 ml of acetic acid solution to prepare β-chitosan solution10.0 ml with sulfuric acid and NN'-dicyclohexylcarbodiimide(DCCI) was added dropwise to the above viscous β-chitosan solution at $0\,°C$, and stirred for 30 mins. The mixed solution was then stirring until to 25 $°C$ and the more stirred for 6 h. The solution was diluted with 100 ml RO water and neutralized by adding 20% NaOH(aq). After neutralization, the solution was precipitated by adding 500 ml MeOH, followed by centrifugation and was dialyzed with RO water. The dried product was obtained by CaCl$_2$ contained vacuum oven.

Finally, the product was centrifuged, and filtering and It's solid gel product was hydrolysis by 20% NaOH 30 ml was added at 80 $°C$ and stirred for 40min. After stirring for 30 min at 25$°C$. and neutralization the solution was precipitated by adding 10% HCl solution and 800 ml of MeOH, followed by centrifugation and was dialyzed with H$_2$O.

Sulfonation Procedure of β-chitin Derivatives

2,3-dihydroxyl-β-Chitin powder (183.2K) 1.5 g was dissolved in 60 ml DMF: 2ml acetic acid solution to prepare tetrasulfonyl-β-chitosan solution. 10.0ml with sulfuric acid and NN'-dicyclohexylcarbodiimide(DCCI) was added dropwise to the above viscous chitosan solution at 0$°C$, and stirred for 40 mins. The mixed solution was then warmed up to 25 $°C$ and stirred for another 12 h. The solution was diluted with 100 ml RO water and neutralized by adding 10% NaOH(aq). After neutralization the solution was precipitated by adding 500 ml MeOH, followed by centrifugation and was dialyzed with RO water. The dried product was obtained by lyophilzation. a,β-dihydroxy - β-Chitosan. Molecular weight of chitosan derivatives were determined by GPC HITACHI L-7110, equipped with TSKGEL G3000PW column. The detection was carried out with a HITACHI RI L-7490 detector. The elution was carried out with 0.2M CH_3COOH and 0.3M CH_3COONa buffer at pH 4.85. Calibration of column was carried out by the use of Shodex Standard P-82 kit with molecular weight between 800 and 5.8 kDa.

Antibacterial Test

Psedomonas areuginosa ATCC 25923 and *Staphylococcus aureus* ATCC 9027 were purchased from the Culture Collection and Research Center (Jeil pharmaceutical, Korea). All these bacteria were inoculated in 100 ml nutrient broth (NB, Merck) and incubated all 37 °C for 19 h. Various deacetylated chitosaneous derivatives and chitosan were added into 200 ml volume flask containing 100 ml KH_2PO_4 (6.25×10^{-4} mol/L)buffer. 1 ml (about 10^9 CFU/ml) of each bacterium to be tested was inoculated in the flasks. Incubation was performed by shaking flask at 150 rpm, 37 °C for 1 h. 0.1 ml of decimal dilutions of samples were spared on nutrient agar plates (*Staphylococcus aureus*) and cysine repiticase agar (*Psedomonas areuginosa*) for colony counting. The inhibitory effect was calculated according to Rito -Munoz and Davidson as follows:

% inhibition= (1-T/C) 100%, C=log CFU/ml of control and T= log CFU/mol of sample

The Water Solubility (WS) of Sulfated-chitosan was measured by methods such as Ryu, etc. To begin with, three sheets of film for measuring water solubility were taken. Then, they were desiccated in the desiccators of 105°C for 24 h to measure the initial content of dry sample. Three sheets of film for measuring water solubility were separately taken and put into the beaker of 50 ml with distilled water of 30 ml. Sealing up the entrance of the beaker by parafilm, we put the beaker in the machine of normal temperature at 25°C. It was agitated sometimes and stored for 24 h. After 24 h, the film which was not resolved in the water was taken out to be desiccated by desiccators for 24 h to measure the content of dry sample. The water solubility of film was indicated by the percentage of quantity resolved in the water against the initial dry sample. The water solubility of each film was measured in 3 repetitious experiments to get average value.

Results and Discussion

The IR spectrum of N-(2,3-disulfonatepropyl)chitosan showed the presence of a carbonyl (1738 Cm^{-1}), SO_3- asymmetric stretching (1267 (Cm^{-1}), and O=S=O symmetric stretching multi peak (1027 Cm^{-1}). And, The IR spectrum of compound (3) showed the presence of

(Cm^{-1}) at: 3443-3200(OH), 3240(NH), 2983-2856(CH), 1450-1420(CH$_2$), 1368, 1062. The ^{1}H-NMR Spectrum also showed the existence of compound(2) δ: 1.46(d, 2H, -CH$_2$-), 1.84(t, 1H, =CH-), 2.7(d, 2H, -CH$_2$-), 3.68(d, 2H, -CH$_2$-), 3.8-4.1 (H2, H3, H4, H5, H6)

It was noticed that solubility of SMAC, and SC increased significantly as compared with that of unmodified β-chitosan. SMAC showed excellent inhibitory result for *Staphylococcus aeruginosa and Pseudomonas aeruginosa* than that of chitosan. In addition, SMAC showed better inhibitory data for *Pseudomonas aeruginosa* than that of SC. due to sulfur content of sulfonation. The above antimicrobial data suggested that substation ratio of sulfonation seemed to play importance role in microbial inhibition toward *Staphylococcus aeruginosa* and *Pseudomonas aeruginosa.*

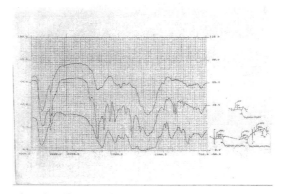

Fig. 1. IR spectrum of tetri-sulfonate polymer chitosan.

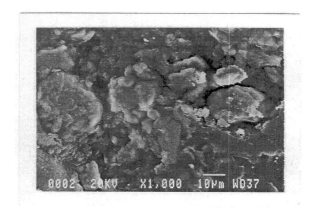

Fig. 2. SEM photograph of sulfonate β-chitosan derivative; surface(X1,000).

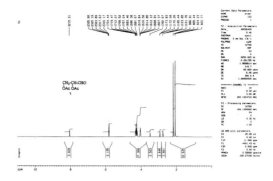

Fig. 3. IR spectrum of 2,3-diacetatylpropionic aldehyde.

Table 1. Antimicrobial Test on Chitosan Derivatives

ATCC is an abbreviation for American Type Culture Collection.

Control		β -Chitosan	SC	SMAC
Staphylococcus aureus ATCC 25923	Inhibitory	64.3	8.32	-
(%)	pH	7.5	7.8	7.5 -
Pseudomonas aeruginosa ATCC 9027	Inhibitor	83	7.1	94
(%)	pH	7.5	8.1	7.4
Escherichia coli ATCC 10536	Inhibitor	93	89	87
(%)	pH	7.5	7.9	7.43

% inhibition= (1-T/C) 100%, C=log CFU/ml of control and T= log CFU/mol of sample

Conclusions

In order to check the appropriateness for skin and pharmacological efficacy, chitosan was converted into sulfate. To be used as biological adhesive, α, β-dihydroxypropionicaldehyde was converted into sulfate by synthesizing 2 with acrolein and the following result came out.

(1) Increment of water solubility. (2) In the reaction of sulfate, the reaction of exothermic sulfuric acid was desirable and the color varied considerably due to the decomposition by temperature.

(3) For H_2SO_4: DCCI: CTS= 2: 10: 1(mole ratio), the temperature under 5°C was desirable.

(4) It seems necessary to check the behavior of electromagnetic wave scattering or gelatin in order to obtain the physical constant that is minutely comparable.

[1]   K. Kurita, M. Kanari, Y. Koyama, *Polym.* **1985**, *14*: 511.
[2]   J. Rudall, J. Thampson, L. Nagy, *J, Vary. Lipids* **1983**, *18*, 714.
[3]   K. Kurita, K. Tomita, *Polym. Bull.* **1993**, *30*, 429
[4]   M. Shimojoh, K. Masaki, K. Kurita, *Nippon Nogeikagaku Kaishi* **1996**, *70*, 787.
[5]   S. ToKura, S. Baba, Y. Uraki, Y. Miura, N. Nishi, O. Hasegawa, *Carbohydrate Polymers* **1990**. *13*, 273.
[6]   S. Hirano, *Carbohydrate. Res* **1976**, *47*, 315.
[7]   M. Shimojo, K. Masaki, K. Kurita, K. Fukushime, *Nippon Nogeikagaku Kaishi* **1996**, *70*, 787.

Macromol. Symp. **2004**, *216*, 55-64

Styrene/(Styrene Derivative) and Styrene/(1-Alkene) Copolymerization using Ph₂Zn-Additive Initiator Systems

Styrene/(Styrene Derivative) and Styrene/(1-Alkene) Copolymerization using Ph$_2$Zn-Additive Initiator Systems

Franco M. Rabagliati,* Rodrigo A. Cancino, Mónica A. Pérez, Francisco J. Rodríguez

Grupo Polímeros, Departamento Ciencias del Ambiente, Facultad de Química y Biología. Universidad de Santiago de Chile, USACH. Casilla 40, Correo 33, Santiago, Chile
E-mail: frabagli@lauca.usach.cl

Summary: Diphenylzinc-metallocene-MAO initiator systems have proven to be effective initiator systems for styrene and for substituted styrenes as well as for their styrene/(styrene-derivative) copolymerization. Titanocene produced almost pure syndiotactic polymers while zirconocenes gave atactic polystyrene together with a low content, less than 20%, of syndiotactic polystyrene. Systems including a zirconocene, particularly ethenyl(bisindenyl)zirconium dichloride were effective initiators of 1-alkene polymerization and of styrene/1-alkene copolymerization. Conversion to polymer increases with the molecular size of 1-alkene. Styrene derivative and styrene/(styrene derivative) polymerization was greatly influenced by the inductive effect of substituent and by steric hindrance due to the monomer.

Keywords: diphenylzinc; metallocene catalysts; styrene polymerization; tacticity

Introduction

Commercial polystyrene, an amorphous polymer, has very reliable properties: it is a good electrical insulator, has excellent optical clarity, and is easy to process making it a suitable material for many and diverse applications However, it is brittle, has poor impact resistance, has a low upper temperature limit, poor weatherability, and is attacked by organic solvents. With the aim to overcome these deficiencies, effort have been made through copolymerization with other monomers, blending with other polymers, and stereoregular polymerization. Both isotactic[1] and syndiotactic [2-4] polystyrene, with Tm = 240 and 270°C, respectively, have been obtained, raising considerably the ceiling temperature of amorphous polystyrene, which is limited by its Tg of 100°C. Isotactic polystyrene, i-PS, has a very low crystallization rate which make it of no commercial interest, while syndiotactic polystyrene, s-PS, has a fast crystallization rate. Beside its high melting temperature, s-PS has high crystallinity and high solvent and thermal resistance. However, due to its rigidity, s-PS is still a brittle material with low impact resistance, and its processing requires a rather high temperature.

DOI: 10.1002/masy.200451207

With the aim of overcoming the deficiencies of s-PS, several research groups have been working on stereoregular styrene polymerization of styrene and on its copolymerization with various monomers. Since the discovery of s-PS in 1985, many papers have been published regarding s-PS synthesis using combined systems including a metallocene and methylaluminoxane, MAO.[5-9] Zambelli et al.[10] have postulated that Ti^{3+} is responsible for syndiotacticity when styrene is polymerized using titanocene/MAO initiators. Copolymerization of styrene with substituted styrenes and of styrene with 1-alkene have also been extensively studied. Po and Cardi published a review on homo- and copolymerization of styrene covering work from the discovery of s-PS until 1994.[11] A more recent review dealing with syndiotactic polystyrene catalysts and polymerization has been published by Schellenberg and Tomotsu.[12]

We have been working on styrene polymerization,[13-17] and on styrene copolymerization including p-substituted styrenes,[18-22,24] and α-olefins. [19,22] We reported that Ph2Zn-metallocene-MAO systems are effective initiators for the homopolymerization of styrene, and depending on the inclusion of a titanocene or a zirconocene, for copolymerization of styrene with substituted styrenes or with α-olefins. According to our results the effectiveness of the initiator system is highly dependent on the metallocene, the styrene-comonomer couple, and the styrene/comonomer mole ratio in the initial feed. The polarity of reaction media also influences the effectiveness of the homopolymerization of styrene and the copolymerization of S/p-alkylstyrene.[17]

In this paper further results on styrene/substituted-styrene and styrene/α-olefin copolymerization are reported, with the purpose to providing additional information to our previous findings regarding the polymerization process, and also with the aim to improving properties of s-PS through S/comonomer copolymerization.

Experimental

Homo- and copolymerization experiments were carried out under argon atmosphere in a 100 cm^3 Schlenk tube equipped with a magnetic stirrer. Solvent toluene (ca., 10 to 30 cm^3), MAO solution, Ph2Zn, and metallocene toluene-solution, were sequentially charged by syringe under argon pressure. Polymerization was initiated by injecting the styrene or simultaneously the required amount of styrene and the second comonomer. The reactions were kept at 60°C under stirring for the required length of time. Polymerization was finished by adding a mixture of hydrochloric acid and methanol. The polymers, coagulated in the acidified methanol, were recovered by filtration after washing several times with methanol, and dried in vacuum at 60°C.

Viscosities were measured either in chloroform or in o-dichlorobenzene depending on the solubility of the polymer and intrinsic viscosities were determined by the one-point method.[19)] For chloroform soluble polymer, viscosity measurements were carried out in chloroform at 25°C and viscosity-average molecular weights, M_v, for a-PS, were calculated according to equation,[20] $[\eta] = 1.12 \times 10^{-4} M_v^{0.73}$, which is reported to be valid for the $7\text{-}150 \times 10^4$ molecular weight range. For s-PS fractions and insoluble in chloroform copolymers, viscosities were measured in o-dichlorobenzene at 135°C.

DSC analyses were performed by using a Rheometrics Scientific DSC apparatus with samples placed under a nitrogen atmosphere. 3 to 4 mg samples were heated at a rate of 10°C/min, and after cooling to room temperature, reheated at the same rate. The reported Tg and Tm were those obtained in the second scan.

NMR spectra were recorded on a Bruker AMX-300 spectrometer at 70°C, operating at 300.1 and 75.5 MHz for ^1H and ^{13}C respectively. The polymers and copolymers were dissolved in deuterated benzene (C_6D_6, 5% w/v). A total of 64 and 4000 scans with 16K and 32K data points and with a relaxation delay of 1 and 2 seconds were collected for ^1H and ^{13}C respectively. Chemical shifts were calibrated to tetramethylsilane (TMS) used as internal reference.

Results and Discussion

For the homopolymerization of styrene we have employed various metallocene combined with MAO and diphenylzinc, Ph_2Zn. In our experiments we have tried both binary, metallocene-MAO, and ternary, Ph_2Zn-metallocene-MAO, initiator systems. Table 1 shows the results of polymerization in which s-PS was obtained when a titanocene was included in the initiator system, while the use of a zirconocene produced atactic polystyrene, a-PS. Anyhow, the crude polystyrene obtained with Ph_2Zn-zirconocene-MAO systems included some s-PS content, less than 20%, established by fractionation in boiling butanone and by NMR analysis. Anyhow DSC thermograms did not show any Tm signal for the PS obtained when using a zirconocene (Figure 1).

As can be seen in Table 1, binary or ternary initiator systems including a titanocene produced s-PS while with zirconocenes or the hafnocene, $(n\text{-}Bu)_2HfCl_2$ mainly a-PS was obtained. We find that efficiency of the initiator systems follows the sequence titanocene > zirconocene > hafnocene.

We have already reported that conversion to polymer increases with p-alkylstyrenes compared with styrene itself. On the other hand, for S/p-halostyrenes copolymerization a substantial decrease in conversion was found, indicating that styrene derivative monomers including I-substituent groups were less reactive than styrene.

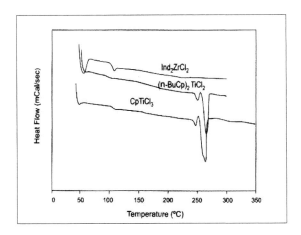

Figure 1. DSC thermograms of crude PS obtained using Ph_2Zn-metallocene-MAO initiator systems in toluene at 60°C. For Ind_2ZrCl_2 and $(n\text{-}BuCp)_2TiCl_2$ metallocenes, 48-hours polymerization. For $CpTiCl_3$, 6-hours polymerization. Second heating at 10°C/min. From Ref.: 22.

Figure 2, shows the results obtained for S/halostyrene copolymerization with the p-chloro-, p-fluoro- and p-bromostyrene. Almost no homo poly(p-halostyrene) was obtained after one hour of polymerization. These results support our previous statement regarding the influence of substitutuents in styrene in terms of copolymer conversion and of the cationic character of the propagation stage. Homopolymerization of p-halostyrene yielded very small amounts of each poly(p-halostyrene).

Table 1. Styrene polymerization using metallocene-MAO and Ph_2Zn-metallocene-MAO initiator systems, in toluene after 48 hours at 60°C.[a]

Metallocene	Metallocene-MAO		Ph_2Zn-Metallocene-MAO	
	Convn. %	Tacticity	Convn. %	Tacticity
Cp_2ZrCl_2	1.0	Atactic	1.1	Atactic
$(i\text{-}BuCp)_2ZrCl_2$	2.5	Atactic	2.7	Atactic
Ind_2ZrCl_2	7.8	Atactic	8.4	Atactic
$(H_4\text{-}Ind)_2ZrCl_2$	1.6	Atactic	2.1	Atactic
$Et(Ind)_2ZrCl_2$	3.1	Atactic	3.3	Atactic
$i\text{-}Pr(Flu)(Cp)ZrCl_2$	8.2	Atactic	8.5	Atactic
$(n\text{-}BuCp)_2HfCl_2$	n.p.[c]	--------	0.6^b	Atactic
Cp_2TiCl_2	1.8	Syndiotactic	6.7	Syndiotactic
$(n\text{-}BuCp)_2TiCl_2$	11.0	Syndiotactic	17.4	Syndiotactic
$CpTiCl_3$	30.5^b	Syndiotactic	42.0^b	Syndiotactic

[a] Polymerization conditions: Total volume = 40 mL, [S] = 2.0 mol/LO, [MAO] = 0.33 mol/L; [metallocene] = 4.0E-04 mol/L for metallocene-MAO systems; $[Ph_2Zn]$ = [metallocene] = 2.0E-04 mol/L for Ph_2Zn-metallocene-MAO systems. [b] After 6 hours polymerization period. [c] not performed.

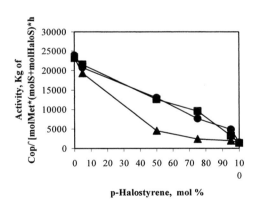

Figure 2. S/p-halostyrene copolymerization using Ph_2Zn-$CpTiCl_3$-MAO initiator system in toluene after 1 hour at 60°C, (●) p-chlorostyrene, (■) p-fluorostyrene, (▲) p-bromostyrene.

In our works we have also tested other substituted styrenes, both on the benzene ring and on the vinyl group. For styrene/α-methylstyrene, S/α-MeS, copolymerization it was found that conversion to polymer decreases as the proportion of α-MeS in the initial feed increases.

Table 2 shows the results obtained for S/α-MeS copolymerization using Ph$_2$Zn-metallocene-MAO initiator systems. Both metallocenes, CpTiCl$_3$ and (n-BuCp)$_2$TiCl$_2$, gave lower conversion to copolymer as the proportion of α-MeS in the initial feed increased. We think that these results are in accordance with a steric hindrance effect at the vinyl bond of the monomer, making coordination of the monomer with the active species more difficult.

Table 2. S/α-MeS copolymerization initiated by Ph$_2$Zn-metallocene-MAO in toluene after 6 hours at 60°C.[a]

Initial Feed S/α-MeS	CpTiCl$_3$			(n-BuCp)$_2$TiCl$_2$		
	Convn.	Tg	Tm	Convn.	Tg	Tm
mol/mol	%	°C	°C	%	°C	°C
S only	59.9	97.8	258.5	10.1	87.8	260.6
95/5	51.7	101.7	257.9	8.3	91.3	263.0
85/15	42.4	98.7	255.9	n.p.	------	------
75/25	34.6	97.8	253.1	1.3	n.p.	n.p.
50/50	23.9	96.1	240.1	0.5	n.p.	n.p.
25/75	6.3	n.p.[b]	n.p.	0.8	n.p.	n.p.
5/95	0.3	n.p.	n.p.	0.6	n.p.	n.p.
α-MeS	none	------	------	0.3	n.p.	n.p.

[a] Polymerization conditions: Total volume = 25 mL, [S] + [α-MeS] = 2.0 mol/L, [MAO] = 0.33 mol/L, [metallocene] = 2.0E-04 mol/L. [b] n.p. = not performed

Figure 3, shows the results of S/substituted styrene copolymerization initiated by the Ph$_2$Zn-CpTiCl$_3$-MAO system for comonomers: 2,4-dimethylstyrene, 2,4-Me$_2$S, 2,4,6-trimethylstyrene, 2,4,6-Me$_3$S, α-methylstyrene, α-MeS. For comparison, the results of S/p-MeS copolymerization are also included. From these results it is seen that crowded comonomers are less prone to being incorporated into the copolymer, and conversion to copolymer decreases. These results suggest that steric effects, and not only electric inductive effects, are important in terms of the efficiency of the initiator systems under study.

These results indicate that not only the electric inductive effects influence the copolymerization process of S/substituted-styrene using Ph_2Zn-metallocene-MAO initiator systems, but also steric hindrance by the incoming monomer has a substantial role in the process.

For styrene/α-olefin copolymerization using Ph_2Zn-metallocene-MAO initiator systems we have used various metallocenes, including titanocenes and zirconocenes. Initiator titanocene systems produced almost pure syndiotactic polystyrene regardless of the S/α-olefin molar ratio in the initial feed. On the other hand, initiator systems including a zirconocene, and particularly $Et(Ind)_2ZrCl_2$, were able to produce true S/α-olefin copolymer.

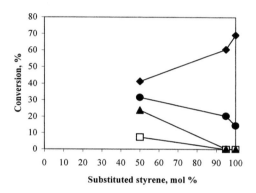

Figure 3. Copolymerization of S/substituted styrene, at various molar ratio S/SubtdS, using Ph_2Zn-CpTiCl$_3$-MAO initiator system in toluene after 6 hours at 60°C. (♦) p-methylstyrene, (●) 2,4-dimethylstyrene, (□) 2,4,6-trimethylstyrene, (▲) α-methylstyrene.

Figure 4 shows the results of α-olefin homopolymerization using Ph_2Zn-$Et(Ind)_2ZrCl_2$-MAO as initiator system. 1-C_nH_{2n} with n = 6, 8, 10, 12, 16 and 18 were tested. Conversion to polymer increases with molecular size of olefin. This result suggest that the length of the hydrocarbon chain of the α-olefin affects the efficiency of the initiator system. The higher conversions were obtained for the homopolymerization of 1-hexadecene and for 1-octadecene, with more than 80% conversion after only three hours polymerization.

When the copolymerization of S/α-olefin was carried out using Ph_2Zn-$Et(Ind)_2ZrCl_2$-MAO as initiator system, true S/α-olefin copolymers were obtained. Figure 5 shows the results obtained for S/α-olefin copolymerization initiated by Ph_2Zn-$Et(Ind)_2ZrCl_2$- MAO initiator system. The highest conversion was obtained for the S/1-C$_{18}$H$_{36}$ couple, confirming that the molecular size of α-olefin has a marked role in S/α-olefin copolymerization when using Ph$_2$Zn-Et(Ind)$_2$ZrCl$_2$-MAO as initiator system.

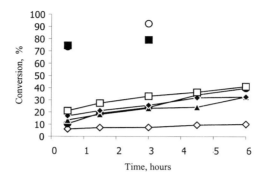

Figure 4. Homopolymerization of α-olefins using Ph$_2$Zn-Et(Ind)$_2$ZrCl$_2$- MAO initiator system in toluene, after various reaction times, at 60°C. (◊) styrene, (♦) 1-C$_6$H$_{12}$, (▲) 1-C$_8$H$_{16}$, (●) 1-C$_{10}$H$_{20}$, (□) 1-C$_{12}$H$_{24}$, (O) 1-C$_{16}$H$_{32}$, (■) 1-C$_{18}$H$_{36}$.

Conclusions

From these and previous results we can conclude that combined systems including diphenylzinc, a metallocene, and methylaluminoxane, Ph$_2$Zn-metallocene-MAO, depending on the metallocene included, induce the homopolymerization of styrene, substituted styrenes and of α-olefins, as well the copolymerization of S/substituted styrenes and of S/α-olefin.

For substituted styrene, conversion to homo- and copolymer with styrene depends largely on the inductive effect of substituent group and on the steric hindrance due to the incoming monomer.

Ph$_2$Zn-zirconocene-MAO systems, particularly when including Et(Ind)$_2$ZrCl$_2$, are effective initiators of homopolymerization of α-olefins and of their copolymerization with styrene.

Conversion to polymer is influenced by the molecular size of the α-olefin, increasing as the length of the olefin's hydrocarbon chain increases.

Further experimental work is in progress from which conclusive results are expected.

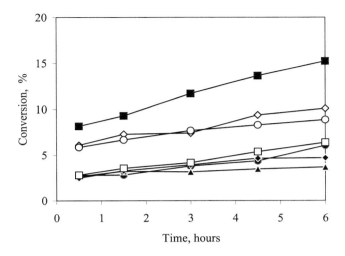

Figure 5. Copolymerization of S/α-olefin (50/50, mol/mol) using Ph_2Zn-$Et(Ind)_2ZrCl_2$-MAO initiator system in toluene, after various reaction times at 60°C. ($\Diamond$) styrene, ($\blacklozenge$) 1-C_6H_{12}, ($\blacktriangle$) 1-C_8H_{16}, ($\bullet$) 1-$C_{10}H_{20}$, ($\square$) 1-$C_{12}H_{24}$, (O) 1-$C_{16}H_{32}$, ($\blacksquare$) 1-$C_{18}H_{36}$.

Acknowledgements

Financial support from Departamento de Investigaciones Científicas y Tecnológicas, Universidad de Santiago de Chile, DICYT-USACH, and from Fondo Nacional de Desarrollo Científico y Tecnológico, FONDECYT, Grant 101-0036, are gratefully acknowledged. The authors also thank Maricel P. Cerda for technical assistance.

[1] G. Natta, P. Pino, P. Corradini, F. Danusso, E. Mantica, G. Mazzanti, G. Moraglio, *J. Am. Chem. Soc.* **1955**, *77*, 1708; G. Natta, P. Pino, G. Mazzanti, *Chimica et Industria*, **1955**, 37, 927.
[2] N. Ishihara, T. Seimiya, M. Kuramoto, M. Uoi, *Macromolecules* **1986**,*19*, 2464.
[3] C. Pellecchia, P. Longo, A. Grassi, P. Ammendola, A. Zambelli, *Makromol. Chem. Rapid Commun.* **1987**, *8*, 277.
[4] W. Kaminsky, *Catalysis Today* **2000**, *62*, 23.
[5] C. Schwecke, W. Kaminsky, *Macromol Rapid Commun* **2001**, *22*, 508.
[6] T. E. Ready, J. W. C. Chien, M. D. Rausch, *J. Organomet. Chem.* **1996**, *Xx*, 21.
[7] W. Kaminsky, S. Lenk, V. Scholz, H. W. Roesky, A. Herzog, *Macromolecules* **1997**, *30*, 7647.
[8] N. Tomutso, N. Ishihara, T. H. Newman, M. T. Malanga, *J. Mol. Cat. A, Chem.* **1998**, *128*, 167.
[9] K. Soga, H. Nakatani, T. Monoi, *Macromolecules* **1990**, *23*, 953.
[10] A. Zambelli, C. Pellecchia, in "Catalyst Design for Tailor-Made Polyolefins". Kodansha(Tokyo)-Elsevier, K. Soga, M. Terano, Eds, **1994**, 209.
[11] R. Po, N. Cardi, *Prog. Polym. Sci.* **1996**, *21*, 47.
[12] J. Schellenberg, N. Tomotsu, *Prog. Polym. Sci.***2002**, *27*, 1925.
[13] F. M. Rabagliati, R. Quijada, M. V. Cuevas, C. Terraza, *Polym. Bull.* **1996**, *37*, 13.
[14] F. M. Rabagliati, C. A. Terraza, R. Quijada, *Intern. J. Polymeric Mater.* **1996**, *34*, 163.
[15] F. M. Rabagliati, M. Pérez, H. A. Ayal, C. A. Terraza, R. Quijada, *Polym. Bull.* **1997**, *39*, 693.
[16] F. M. Rabagliati, M. A. Pérez, R. Quijada, *Polym. Bull.* **1998**, *41*, 441.

[17]   F.M. Rabagliati, M.A. Pérez, R. Cancino, R. Quijada R., *Polym. Int.* **1999**, *48*, 681.

[18]   F.M. Rabagliati, M.A. Pérez, M.A. Soto, A. Matínez de Ilarduya, S. Muñoz-Guerra, *Eur. Polym.J.* **2001**, *37*, 1001.

[19]   F.M. Rabagliati, M.A. Pérez, R. Cancino, M.A. Soto, F.J. Rodríguez, C.J. Caro, A.G. León,H.A. Ayal, R. Quijada, *Macromol. Symp.* **2001**, *168*, 31.

[20]   F.M. Rabagliati, J.C. Caro, M.A. Pérez, *Bol. Soc. Chil. Quím.* **2002**, *47*, 137.

[21]   F.M. Rabagliati, M.A. Pérez, R.A. Cancino, M.A. Soto, F.J. Rodríguez, C.J. Caro, *Macromol.Symp.* **2003**, *195*, 81.

[22]   F.M. Rabagliati, M.A. Pérez, R. Cancino, M.A. Soto, F.J. Rodríguez, A.G. León, H.A. Ayal, R.Quijada, *Bol. Soc. Chil. Quím.* **2000**, *45*, 219.

[23]   F.M. Rabagliati, R.A. Cancino, F.J. Rodríguez, *Polym. Bull.* **2001**, *46*, 427.

[24]   F.M. Rabagliati, M.A. Pérez, F.J. Rodríguez, C.J. Caro, N. Crispel, *Polym. Int*, submitted.

Macromol. Symp. **2004**, *216*, 65-76   65

Functional Water Soluble Polymers with Ability to Bind Metal Ions

Bernabé L. Rivas,* Eduardo Pereira

Faculty of Chemistry, University of Concepción, Casilla 160-C, Concepción, Chile
E-mail: brivas@udec.cl

Summary: Water-soluble polymers containing amine, carboxylic acid, and sulfonic acid groups were investigated as polychelatogens through the liquid phase polymer-based retention, LPR technique, under different experimental conditions. The metal ions investigated are: Ag(I), Cu(II), Co(II), Ni(II), Ca(II), Hg(II), and Cr(III). An important effect of the pH and the ligand type was observed on the metal ion retention. As the pH increases the metal ion retention increases. Two types of metal ion interactions are involved: coordination and electrostatic.

Keywords: functionalization of polymers; water-soluble polymers

Introduction

Water-soluble polymers (WSP) containing ligands at the main or side chains have been investigated for the removal of metal ions in the homogeneous phase.[1-6] These chelating polymers, termed *polychelatogens*, have been prepared by functionalizing various basic polymers. Water-soluble polymers are commercially available or can be synthesized by different routes. Among the most important requirements for technological applications of these polymers, are their high solubility in water, easy and cheap route of synthesis, an adequate molecular weight and molecular weight distribution, chemical stability, high affinity for one or more metal ions, and selectivity for the metal ion of interest. The macromolecules can be homo- or copolymers, and may contain one or more coordinating and/or charged groups. These groups are placed at the backbone, or at the side chain, directly or through a spacer group. Using these polychelatogens in combination with membrane filtration, higher efficiency and selectivity of membrane separations can be achieved.

Polyelectrolytes may be distinguished from chelating polymers (polychelatogens). The former have charged groups, or easily ionizable groups in aqueous solution, while the latter bears

              DOI: 10.1002/masy.200451208

functional groups with the ability to form coordination bonds. The most investigated ligands present in the polychelatogens are amines, carboxylic acids, amides, alcohols, aminoacids, pyridines, thioureas, iminos, *etc.* Among them, polymers containing amino groups have been extensively studied by ultrafiltration, particularly the functional poly(ethyleneimine). This heterochain polymer contains three different types of amino groups: secondary and tertiary groups in the main chain and secondary and primary amino groups in the side-chain. The ratios are between 1:1:1 and 1:2:1 for the primary, secondary, and tertiary species, varying depending on the degree of branching. The most studied polyelectrolytes include those with carboxylic acid, phosphoric acid, sulfonic acid, or amonio groups in their structure.

A water-soluble polymer solution may be considered a two phase system. The polymeric functional groups keep separated by a distance contained in a range so that they present a local high and nearly constant concentration. The water-soluble polymers undergo interactions with solvent and other high and low-molecular weight species present in the aqueous solutions. Due to these interactions, macromolecules in solution exhibit, beyond their chemical structure, different properties such as conformation of the polymer chains, excluded volume.

Membrane filtration processes can be successfully used for the separation of inorganic species and for their enrichment from dilute solutions with the aid of a water-soluble polymer. This technique is called the *liquid-phase polymer based retention*, LPR, technique. Ultrafiltration is fast-emerging as a new and versatile technique in separation technology, concentration, purification, and separation processes. Ultrafiltration of water-soluble, high molecular-weight polymers in the presence of low molecular weight electrolytes or molecules, allows the detection of interactions between the macromolecules and the low molecular weight of small species, such as metal ions.

This technique uses solutions of polychelatogens in combination with membrane filtration where selective separations of metal ions can be achieved. When solutions of mixtures of metal ions are put in contact with the polymer solution, a profile for the retention of the different metal ions by the polychelatogen during the filtration, can be obtained.

A series of polymers have been designed and investigated with respect to the analytical determination of metal ions. Poly(ethyleneimine)-based reagents have been used in many studies as complexing ligands, as well as a versatile source of chelating derivatives for removing metal ions from aqueous solutions by means of complexation-ultrafiltration (UF). [4, 5, 7-24] Among these, poly(ethyleneimine)-methyl phosphoric acid was tested as a macromolecular reagent to the isolation and analysis of plutonium in contaminated waters from the area near to the Chernobyl nuclear power plant.[10] Poly(acrylic acid) and copolymers with acrylamides[13-14], N-maleyl glycine [15], N-vinyl-2-pyrrolidone[14], poly(2-acrylamido-2-methyl-1-propane sulfonic acid)[16], poly(N-acryloyl-N-methyl piperazine)[17], as well as mixtures of two polychelatogens [18] were investigated in their ability to bind di- and trivalent cations.

Liquid-Phase Polymer-Based Retention (LPR) Technique

The method, *liquid-phase polymer based retention* (LPR) is based on the retention of certain ions by a membrane which separates low molecular mass compounds from macromolecular complexes of the ions. Thus, uncomplexed inorganic ions can be removed by the filtrate, whereas the water-soluble polymer complexes are retained (see figure 1).

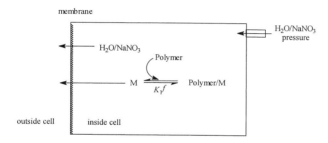

Fig. 1. Ultrafiltration process.

Different modes of separation by LPR can be used for inorganic ions. To separate the components of a small volume sample in analytical chemistry (relative preconcentration), the liquid sample is placed in the polymer containing cell solution and then washed with water (washing method). The pH is adjusted to a value at which the ions of interest are retained and the other species are removed. The washing method can also be applied to purify a macromolecular compound by eliminating the microsolutes maintaining a constant volume in the cell.

To achieve enrichment of the metal ions, their solution can be passed from a reservoir to a smaller volume filtration unit in the presence of a complexing polymer. This concentration method (enrichment method) is designed for metal recovery from dilute technological solutions and for absolute preconcentration of elements in analytical chemistry.

The main features of a *liquid-phase polymer-based retention* system are a membrane filtration, reservoir and a pressure source, *e.g.* a nitrogen bottle. Conventional stirred filtration cells or a specially designed tangential-flow cell equipped with a pump can be used. Essential parameters are the molecular mass exclusion rate in wide pH range (1-12), an appropriate permeate flow rate $(0.5-12 \text{ mL min}^{-1})$, retentate volume (2-50 mL) and gas pressure 300 kPa, are suitable in most cases. A polymer concentration of 0.5-5 weight-% in the cell solution is more appropriate for both retentions of elements and their subsequent determination in the retentate. The most usual molecular mass cut-off ranged between 1000 and 300000 Daltons. A nominal exclusion rate of 10 kg mol^{-1} proved to be convenient for polymers having a molecular mass ranging between 30 and 50 kg mol^{-1}.

The advantages of this method are the use of the separation, owing to the use of a selective binding, and the low energy requirements involved in UF.

The most common ultrafiltration membranes are based on cellulose acetate (CA), polysulfone (PS), poly(ether sulfone) (PES), polyacrylonitrile (PAN) and polyvinyldiene (PVD). Although CA-based membranes are less prone to fouling and typically have a higher flux than PS membranes at equivalent rejection, PS membranes are used in many applications because of higher stability (see table 1).

Table 1. Molecular structure of ultrafiltration membranes.

Name	Molecular structure
Poly(sulfone)	
Cellulose acetate	
Poly(ethersulfone)	
Poly(ethylene)	$-(CH_2)_n-$
Poly(vinyldiene fluoride)	$-(CH_2CF_2)_n-$

Moreover, hydrophobic polymers and membranes can be modified to increase flux and reduce fouling. These membrane modification techniques include the following: entrapping a hydrophilic moiety, reacting the base polymer with hydrophilic pendant groups, blending polymers, modifying the membrane surfaces, modifying the membrane charge and grafting the ceramic membranes.

Ultrafiltration is based on the same principle as reverse osmosis but with much larger pore sizes (0.001-0.1 µm). Because of the low or negligible osmotic pressure of macrosolutes, ultrafiltration operates at very low pressures (30-80 psi).

Ultrafiltration membrane separation depends on membrane pore size, solute-membrane interaction, and shape and size of the macromolecule. For maximum separation efficiency, there should be a 10-fold difference in the sizes of the species in order to be separated. In addition, because all ultrafiltration membranes have a certain pore size distribution, the molecular mass cut off (MMCO) of the membrane (defined as the molecular weight at which 90% of the macromolecular solute is rejected by the membrane) should be at least one-half that of the smallest macromolecule that must be rejected.

In order to quantify the separation process, retention of the metal species is plotted versus the filtration factor. The binding and elution processes may be formulated as a chemical reaction, where reversible reaction in combination with an irreversible transfer of metal ions across the membrane is responsible for metal retention. Retention (R_Z) is defined for any species as the fraction per unit of the species under study remaining in the cell during filtration. The filtration factor (F) is defined as the volume ratio of the filtrate (V^f) versus volume in the cell (V^c). The metal ion (M) remaining in the cell during filtration consists of the sum of the metal ion bound to the polymer chain and the metal ion free in the solution. These values are a function of F, $i.e.$, the extent of the filtration run. Then, when the volume inside the ultrafiltration cell is kept constant during filtration, retention may be formulated as follows:

$$R_Z(F) = \frac{c_Z^{free}(F) + c_Z^{bound}(F)}{c_Z^{init}} \qquad (1)$$

where c_Z^{free} is the concentration of M free in the solution, c_Z^{bound} is the concentration of M bound to the polymer, and c_Z^{init} is the initial metal concentration. Z is the valence of the metal ion considered.

The Nature of the WSP-Metal Ion Interactions

Interactions of metal ions with water-soluble polymers are mainly due to electrostatic forces and the formation of coordinating bonds. Other weak interactions may appear as trapping metal ions in the bulk of the polymer phase.

The coordination features between polymers and metal ions may be described by the usual coordination theories, but some special aspects may be considered. When intrachain complexes are predominant, it is normally recognized the existence of a polymer domain with near constant concentration of ligands, since the distances between them are kept in a narrow range for a given polymer chain. This is responsible that the chelating reaction appears as a one-step reaction. For relative low ligand amounts in the solution, completely complexed metal ions may coexist with free, uncomplexed ones.

Polychelatogens

Poly(ethyleneimine) PE, poly(ethyleneimine-*co*-epichlorohydrine) PEE, poly(methacrylic acid) PMA, poly(2-acrylamido-2-methyl-1-propane sulfonic acid) PAPSA, were synthesized, except the commercial PE and PEE, from the respective monomers by radical polymerization. Subsequently the polymers were purified by dissolving in water and passing through the ultrafiltration membrane with a molecular mass cut-off 3.000 D. The final product was lyophilized.

The structures of the polychelatogens are the following:

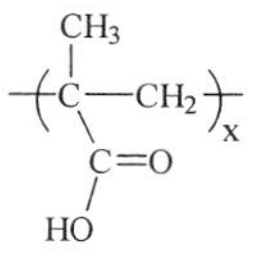

PMA

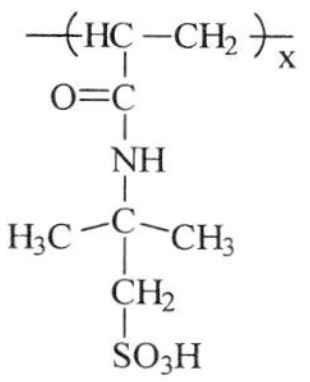

PAPSA

$-(NHCH_2CH_2)_x(NCH_2CH_2)_y-$
$\quad\quad\quad\quad\quad\quad\quad |$
$\quad\quad\quad\quad\quad CH_2CH_2NH_2$

PE

$-(CH_2CH_2-N)_x(CH_2CH_2-N)_y-$
$\quad\quad\quad\quad |$
$\quad\quad RCH_2CHCH_2 \quad\quad (CH_2CH_2NH)_zH$
$\quad\quad\quad\quad |$
$\quad\quad\quad\quad OH$

PEE

Procedure to Study the Metal Ion Binding Properties

The membrane filtration Filtron, poly(ether sulfone)] was carried out in a system described previously. [4,11] For the determination of the complex binding ability, 0.02 M aqueous solution of the polychelatogen was prepared and adjusted to the corresponding pH by addition of diluted HNO_3 or NaOH. The solutions of polymer and metal nitrates (0.0008 M) were placed into the filtration cell. The total volume in the cell was kept constant at 20 mL. The system was pressurized by nitrogen gas and kept constant at 300 kPa during the membrane filtration. The filtration fraction ($Z = 1$-10) was recollected and the concentration of metal ions in the filtrate was determined by atomic absorption spectrometry. Z is defined as the ratio of the volume in the filtrate (V_f) and the volume of the cell solution (V_o).

Polymer-metal Ion Retention Properties

Retention can be plotted versus the filtration factor, Z, and a retention profile is obtained. R is the fraction per unit of metal ions remaining in the cell. As an example, figure 2 shows several retention profiles obtained for different metal ions by PE.

The polymer metal ion interaction is pH-dependent. In strongly acidic solutions no complexation takes place. At pH=3 no metal ion was significantly retained. The highest ability is observed at pH 7 for all the metal ions. At this pH it is very important the strong amine-metal ion interaction as at $Z = 10$, the retention is kept over 80 % for C(II), Cd(II), Cu(II), Ni(II), Zn(II), and Co(II). The PE shows only a poor afinity for Pb(II).

The polychelatogen PEE shows a lower metal ion retention capability than that PE. It is due to that the copolymer includes hydroxyl groups which are more weak ligands compaired with amino groups. As an example, figure 3 shows the retention behavior for Ni(II) by PE and PEE.

The PMA shows also an important effect of the pH on the metal ion retention. At pH = 5, the highest retention (100%) was obtained for Cu(II) ions. Cr(III) ions precipitate at this pH and were not investigated. At pH = 7, the Ag(I), Cd(II), Co(II), Ni(II) were retained in a 100%. At such pH all the carboxylic acids are dissociated and the carboxylate groups form more stable complexes with the divalent cations. Cu(II) precipitates at this pH, hence it was not investigated.

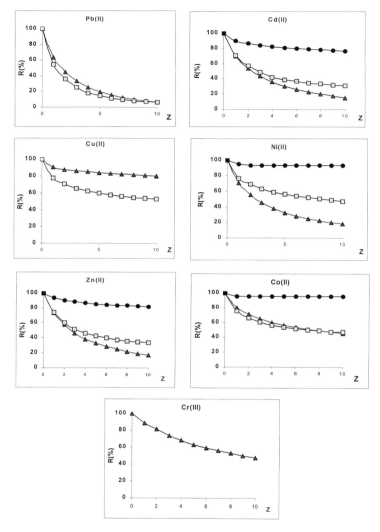

Fig. 2. Retention profiles for poly(ethyleneimine) PE at pH 3; (σ); 5 (ν), and 7 (λ).

On the contrary, the polychelatogen PAPSA shows for all metal ions a high metal ion retention which is not very dependent on the pH.

It is well known that the filtration factor, Z, is a measurement of the strength of the ligand-metal ion interaction. At Z high enough values, a remaining residue of metal ions is frequently found which cannot be eluted by filtration at the same pH and under the same conditions. Therefore, figure 4 shows that the metal ions Ag(I), Ni(II), Co(II), Cd(II), and Zn(II) form very stable complexes with carboxylate groups at pH=5 and pH=7. At pH=3 the strongest complex was formed with Cr(III). Hg(II), Cu(II), and Cr(III) was not investigated in all pH ranges to avoid precipitation.

Figure 5 shows the metal ion binding properties of the polychelatogens containing the three ligand types. These experiences were carried out at pH = 7 corresponding to the maximum retention and Z = 10. PAPSA is the polychelatogen which shows the strongest interactions with the metal ions and the PEE the weaker ligand-metal ion interactions. Cd(II) ion forms very stable complexes with amino groups coming from PE.

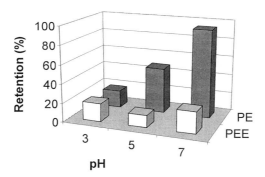

Fig. 3. Retention of Ni(II) by the polychelatogens PE and PEE at different pH.

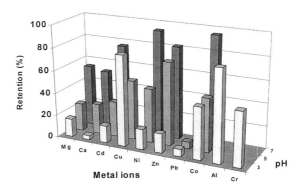

Fig. 4. Metal ion retention of PE at different pH and Z=10.

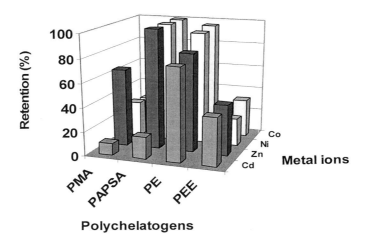

Fig. 5. Metal ion retention for the polychelatogens at Z = 10 and pH = 7.

The capacity of soluble polymer complexing agents is influenced by various factors, mainly steric and statistical ones. Usually there is not only intramolecular, but also intermolecular complexation because the coordination number of metal ions is greater than two. The dependence

of capacity on concentration of the polymer solution differs from polymer to polymer.
Generally, sterical hindrance diminishes the capacity if a certain concentration of solution is exceeded.

Polymer-metal Ion Interaction Mechanism

It is strongly related with the dissociation degree of the polychelatogens. Thus, for PE and PMA, at pH 5 most of the ligand groups are as free amine, and carboxylate respectively. Therefore, these groups form polymer-metal complexes, which are shown in figures 6a) and 6b). APSA is deprotonated at lower pH than that of carboxylic acid. Due to that APSA shows a higher metal ion retention capability. For this polychelatogen, the polymer-metal ion interaction is only an electrostatic type (see figure 6c).

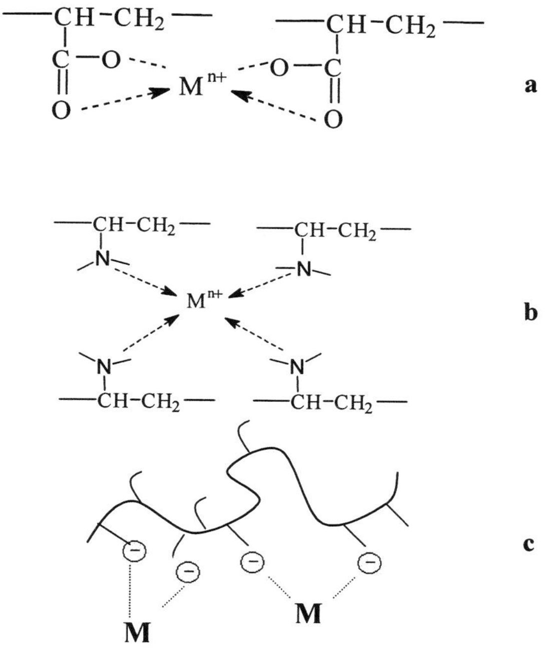

Fig. 6. Possible polymer-metal ion interaction mechanism for a) carboxylate group (complexes formation), b) amine type complexes, and c) sulfonate group (electrostatic interaction).

Acknowledgements

The authors thank to FONDECYT (Grant No 1030669) for the financial support.

[1]   A. D. Pomogailo, G.I. Dzhardimalieva, *Russ. Chem.Bull.* **1998**, *47*, 2319.
[2]   S. K. Chaterjee, A.M. Khan, S. Ghosh, *Angew. Makromol. Chem.* **1992**, *200*, 1.
[3]   E. Tsuchida, K. Abe, *Adv. Polym. Sci.* **1982**, *45*, 1.
[4]   K. E. Geckeler, G. Lange, H. Eberhardt, E. Bayer, *Pure Appl. Chem.* **1980**, *52*, 1883.
[5]   B. L. Rivas, K. E. Geckeler, *Adv. Polym. Sci.* **1992**, *102*, 171.
[6]   R. A. Beauvais, S. D. Alexandratos, *Reactive Polymers* **1998**, *36*, 113.
[7]   K. E. Geckeler, R. Zhou, A. Fink, B. L. Rivas, *J. Appl. Polym. Sci.* **1996**, *60*, 2191.
[8]   K. E. Geckeler, E. Bayer, B.Ya. Spivakov, V. M. Shkinev, G. A. Vorobeva, *Anal. Chim. Acta*, **1986**, *189*, 285.
[9]   K. E. Gekeler, R. Zhou, B. L. Rivas, *Angew. Makromol. Chem.* **1997**, *197*, 102.
[10]   K. E. Geckeler, R. Zhou, *Naturwissenschaften*, **1993**, *75*, 198.
[11]   B. Y. Spivakov, K. E. Geckeler, E. Bayer, *Nature*, **1985**, *315*, 313.
[12]   E. Bayer, H. Eberhardt, P. A. Grathwohl, K. E. Geckeler, *Israel J. Chem.* **1985**, *26*, 40.
[13]   B. L. Rivas, S. A. Pooley, M. Soto, H. A. Maturana, K. E. Geckeler, *J. Appl. Polym. Sci.* **1998**, *67*, 93.
[14]   B. L. Rivas, S. A. Pooley, M. Soto, K. E. Geckeler, *J. Polym.Sci. Part A: Polym. Chem.* **1997**, *35*, 2461.
[15]   G.del. C. Pizarro, O. Marambio, B. L. Rivas, K. E. Geckeler, *J. Macromol. Sci. Pure Appl. Chem.* **1997**, *A34*, 1483.
[16]   B. L. Rivas, I. Moreno-Villoslada, *J. Phys. Chem.* **1998**, *102*, 6994.
[17]   B. L. Rivas, S. A. Pooley, M. Luna, *Macromol. Rapid Commun.* **2000**, *13*, 905.
[18]   B. L. Rivas, I. Moreno-Villoslada, *Macromol. Chem. Phys.* **1998**, 199, 1153.
[19]   S. Masuda, T. Kobayashi, T. Tomida, T. Inoue, M. Tanaka, Y. Asahi, *Polymer* **1993**, *34*, 4562.
[20]   B. L. Rivas, E. D. Pereira, I. Moreno-Villoslada, *Prog. Polym. Sci.* **2003**, *28*, 173.
[21]   B. L. Rivas, L. N. Schiappacasse, *J.Appl.Polym.Sci.* **2003**, *88*, 1698.
[22]   B. L. Rivas, E. D. Pereira, P. Gallegos, K. E. Geckeler, *Polym. Adv. Technol.* **2003**, *13*, 1000.
[23]   I. Moreno-Villoslada, B. L. Rivas, *J. Membrane Sci.* **2003**, *215*, 195.
[24]   B. L. Rivas in "Advanced Supramolecular and Supramolecular Materials and Processes" Editor: Kurt E.Geckeler. Kluver Academic/Plenum Publishers, New York pp 251-264 (**2002**).

Macromol. Symp. **2004**, *216*, 77-85

Approaches to Conjugated Polymers *via* New Solid State Polymerizations

Daniel J. Sandman,* Jeffrey M. Njus, Bao Tran

Center for Advanced Materials, Department of Chemistry, University of Massachusetts Lowell, Lowell, Massachusetts 01854-5046 USA

Summary: Solid state polymerization is the most convenient approach for the synthesis of macroscopic polymer single crystals. Motivations for seeking fully crystalline specimens of conjugated polymers other than polydiacetylenes (PDA) are presented. The thermal reactivity of *p*-ethynylbenzoiic acid (EBA) is found to be a topochemically initiated solid state polymerization. The studies of EBA led to a new paradigm for solid state reactivity in such systems. Systems related to the paradigm, namely, *p*-cyanobenzoic acid (CNBA) and *p*-cyanocinnamic acid (CNCA) have been studied. Sublimation competed with thermal processes in CNBA and CNCA, and a detailed product description has proved difficult to date. Thermochromism and charge-transfer transitions in PDA are discussed.

Keywords: conjugated polymers; monoacetylenes; nitriles; polydiacetylenes; solid state polymerization; thermal reactivity

Introduction

The 2000 Nobel Prize in Chemistry [1-3] awarded to H. Shirakawa, A.G. MacDiarmid, and A.J. Heeger, is a clear recognition of the impact that conjugated polymers have had in the areas of scientific, technological, and commercial activity. Yet, when placed in a broader context of materials, as will be done herein, it is apparent that there is considerable opportunity for further progress in this area of endeavor.

Many classes of materials, but especially inorganic semiconductors, exist in both crystalline and amorphous forms. The disordered structure of the amorphous forms leads to lower carrier mobilities, and the electrical properties of amorphous forms are inferior to those of single crystals.[4] Crystalline silicon is an indirect band gap semiconductor while its amorphous form has a direct gap. The temperature dependence of the conductivity of the two forms is significantly different at low temperatures.[5] Nevertheless, amorphous semiconductors are readily fabricated into large area thin film devices with commercially viable applications such as those known for the amorphous chalcogenides Se and As_2Se_3. The latter have been

DOI: 10.1002/masy.200451209

extensively used in copying machines as photoreceptors.

The majority of the research and successes to date in the area of conjugated polymers have been largely achieved in materials that are amorphous or partially crystalline at best.[1-3] While the need for improved structural order has been emphasized,[3] the existing materials, with the single exception of polydiacetylenes (PDA), are amorphous materials whose fully crystalline analogs are unknown materials whose properties can occasionally be guessed at, for example, by extrapolation of the properties of rigorously defined oligomers.[6] The molecular structures of the PDA under discussion are given in Figure 1. That we are missing substantial information with the lack of numerous fully crystalline conjugated polymers can readily be grasped from the example of the electronic spectrum of a PDA that has both crystalline and amorphous forms, namely PDA-4-BCMU.[7] (Scheme 1, **1a**) With light polarized parallel to the conjugated polymer chain, the crystalline form exhibits an intense spectral maximum at 633 nm with vibronic structure. Perpendicular to the conjugated chain, the spectrum is markedly weaker and featureless, as expected for an anisotropic material. In contrast, the amorphous film exhibits broad maxima at 526 and 488 nm. Clearly, there is markedly less information in the spectrum of the amorphous PDA film relative to that of the single crystal polymer specimen. By extension, there is considerable loss of information in the spectral and other properties of the large number of reported amorphous conjugated polymers.

PDA are a class of fully crystalline conjugated polymers whose modern study was pioneered by G. Wegner and coworkers. By far, PDA constitute the best-known class of fully crystalline conjugated polymers.[8-15] As a consequence of their synthesis by topochemical (atoms making a new covalent bond are separated by about 3.8-4.2Å in the reactant crystal structure) and topotactic solid state polymerization, many PDA are available in the form of macroscopic single crystals, and hence provide several examples of the best defined polymers.[8-15] As such, PDA join the metallic and superconducting sulfur nitride[16] and dihalobis (*tris* 2-cyanoethylphosphine) nickel (II) polymers[17] as examples of crystallographically defined polymers obtained via solid-state polymerization.[10] The diacetylene (DA) polymerization process is more general than the other cases.

While PDA clearly have attractive physical properties, there are good scientific and technological reasons for seeking other examples of fully crystalline conjugated polymers. If one is interested in polymers that exhibit fluorescence and properties related to fluorescence such as electroluminescence, crystalline PDA have very low quantum yields of fluorescence,[13] and other examples must be sought. In addition, approaches to fully

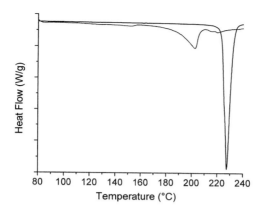

Figure 3. Melting behavior of CNBA by differential scanning calorimetry. The first scan has an endotherm of 220.0 J/gm. While the heat associated with the second scan is 62.6 J/gm.

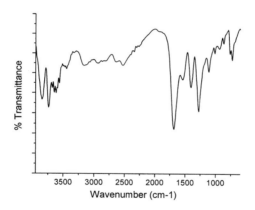

Figure 4. FTIR spectrum of the product of thermal reaction of CNBA.

Polydiacetylene Spectral Features Below the Exciton

The PDA of 1,6-diethyl-(1,6-bis-*p*-benzylidenemalononitrile)-2,4-hexaidyndiamine (**1c**) in single crystal form exhibits[40] heretofore unobserved spectral features in both principal directions from about 833-625 nm. The energy of this band, its width, lack of vibronic structure, and apparent polarization suggest that it arises from one or more charge-transfer transitions. The donor would be the conjugated backbone, and the acceptor would be a side group of an adjacent chain. Subsequently, such absorption was reported[41] in another PDA.

84

Acknowledgements

We thank the Petroleum Research Fund (40263-AC7) for partial support of this work. Bao Tran is a student at Lowell High School. His participation was supported in part by the U.S. Army Research and Engineering Program through the Academy of Applied Sciences, Concord, NH.

[1]   H. Shirakawa, *Angew. Chem. Int. Ed. Eng.* **2001**, *40*, 2574.
[2]   A. G. MacDiarmid, *Angew. Chem. Int. Ed. Eng.* **2001**, *40*, 2581.
[3]   A. J. Heeger, *Angew. Chem. Int. Ed. Eng.* **2001**, *40*, 2591.
[4]   P. A. Cox, *The Electronic Structure and Chemistry of Solids,* Oxford University Press, **1987**, pp. 218-221.
[5]   J. I. Gersten, F. W. Smith, *The Physics and Chemistry of Materials,* John Wiley and Sons, New York, **2001**, pp. 391-392.
[6]   K. Müllen, G. Wegner, eds., *Electronic Materials: The Oligomer Approach,* Wiley-VCH **1998**.
[7]   R. R. Chance, G. N. Patel, J. D. Witt, *J. Chem. Phys.* **1979**, *71*, 206. `
[8]   H. J. Cantow, Ed. *Polydiacetylenes* Advances in Polymer Science, Springer Verlag, Vol. 63, **1984**.
[9]   D. Bloor, R. R. Chance, Eds., *Polydiacetylenes* Martinus Nijhoff, NATO ASI Series, Dordrecht, Boston, **1985**.
[10]   D. J. Sandman, Ed., *Crystallographically Ordered Polymers* American Chemical Society Symposium Series Vol. 337, **1987**.
[11]   R. R. Chance, in *Encyclopedia of Polymer Science and Engineering;* J.I. Kroschwitz, Ed., Wiley-Interscience, 2nd ed., **1986**, vol. 4, pp. 767-779.
[12]   M. Pope, C. E. Swenberg, *Electronic Processes in Organic Crystals;* Oxford University Press, 2nd ed., **1999**, pp. 673-699.
[13]   D. J. Sandman, in: *Polymeric Materials Encyclopedia;* J.C. Salamone, Ed., CRC Press, **1996**, Vol. 2. pp. 1468-1480.
[14]   W. D. Huntsman, in:*The Chemistry of Functional Groups, Supplement C;* S. Patai, Z. Rappoport, Eds., **1983**, Wiley.
[15]   H. Nakanishi, in: *Polymeric Materials Encyclopedia;* J.C. Salamone, Ed., CRC Press, **1996**, vol.10, pp. 8393-8398.
[16]   M. M. Labes, P. Love, L. F. Nichols, *Chem. Rev.* **1979**, *79*, 1.
[17]   K. Cheng, B. M. Foxman, *J. Am. Chem. Soc.* **1977**, *99*, 8102.
[18]   D. J. Sandman, G. P. Hamill, L. A. Samuelson, B. M. Foxman, *Mol. Cryst. Liq. Cryst.* **1984**, *106*, 199.
[19]   D. J. Sandman, C. S. Velazquez, G. P. Hamill, B. M. Foxman, J. M. Preses, R.E. Weston, Jr., *Mol. Cryst. Liq. Cryst.* **1988**, *156*, 103.
[20]   G. Ribera, M. Galià, V. Cádiz, *Macromol. Chem. Phys.* **2001**, *202*, 3363.
[21]   A. P. Mellissaris, M. H. Litt, *J. Org. Chem.* **1992**, *57*, 6998.
[22]   A. P. Mellissaris, M. H. Litt, *Macromolecules* **1994**, *27*, 2675.
[23]   T. Steiner, E. B. Starikov, A. M. Amado, J. J. C. Teixera-Dias, *J. Chem. Soc., Perkin Trans. 2*, **1995**, 1321.
[24]   J. M. A. Robinson, B. M. Kariuki, R. J. Gough, K. D. M. Harris, D. Philip, *J. Solid State Chem.* **1997**, *134*, 203-206.
[25]   P. Cadiot, W. Chodkiewicz, "Coupling of Acetylenes", in *Chemistry of Acetylenes,* H.-G. Viehe, ed., Marcel Dekker, New York, **1969**, pp. 597-647.
[26]   S. A. Swanson, W. W. Fleming, D. C. Hofer, *Macromolecule*s **1992**, *25*, 582.
[27]   E. Weber, M. Hecker, E. Koepp, W. Orlia, M. Czugler, I. Csoregh, *J. Chem. Soc., Perkin Trans. 2*, **1988**, 1251.
[28]   J. Njus, D. J. Sandman, L. Yang, B. M. Foxman, *Polymer Preprints* **2003**, *44(1)*, 905.
[29]   A. M. Cianciusi, A. Furlani, A. L. Ginestra, M. V. Russo, G. Palyi, A. Visi-Orosz,. *Polymer* **1990**, *31*, 1569.
[30]   Aldrich Handbook of Fine Chemicals and Laboratory Equipment, **2003-2004** edition, p. 518.
[31]   E. P. Valby, H. J. Lucas, *J. Am. Chem. Soc.* **1929**, *51*, 2718.
[32]   T. Higashi, K. Osaki, *Acta Cryst.* **1981**, *B37*, 777.
[33]   N. Moses, *Ber. Deut. Chem. Gesell.* **1900**, *33*, 2625.

[34]   M. S. K. Dhurjati, J. A. R. P. Sarma, G. R. Desiraju, *J. Chem. Soc., Chem. Commun.* **1991**, 1702.
[35]   R. R. Chance, R. H. Baughman, H. Müller, C. J. Eckhardt, *J. Chem. Phys.* **1977**, *67*, 3616.
[36]   D. J. Sandman, *Trends Polymer Sci.* **1994**, *2*, 44.
[37]   D.-C. Lee, S.K. Sahoo, A. L. Cholli, D. J. Sandman, *Macromolecules* **2002**, *35*, 4347.
[38.   M. J. Downey, G. P. Hamill, M. Rubner, D. J. Sandman, C. S. Velazquez, *Die Makromol. Chem.* **1988**, *188*, 1199.
[39]   V. Enkelmann, Max-Planck-Institut für Polymerforschung, Mainz, private communication.
[40]   J. L. Foley, L. Li, D. J. Sandman, M. J. Vela, B. M. Foxman, R. Albro, C. J. Eckhardt, *J. Am. Chem. Soc.* **1999**, *121*, 7262.
[41]   J. M. Pigos, Z. Zhu, J. Musfeldt, *Chem. Mater.* **1999**, *11*, 3275.

Macromol. Symp. **2004**, *216*, 87-97

Acid-Containing Tyrosine-Derived Polycarbonates: Wettability and Surface Reactivity

Vipavee P. Hoven,[1] *Adisorn Poopattanapong,*[2] *Joachim Kohn*[3]

[1] Organic Synthesis Research Unit, Department of Chemistry, Faculty of Science, Chulalongkorn University, Bangkok 10330, Thailand
E-mail: vipavee.p@chula.ac.th
[2] Program of Petrochemistry and Polymer Science, Faculty of Science, Chulalongkorn University, Bangkok 10330, Thailand
[3] Department of Chemistry, Rutgers University, Piscataway, New Jersey 08855-0939, USA

Summary: Tyrosine-derived polycarbonates having carboxylic acid pendant groups were characterized by water contact angle and X-ray photoelectron spectroscopy (XPS). A pronounce decrease of receding angle as well as contact angle hysteresis as a function of acid composition strongly indicated that the acid groups are more accessible at the water/polymer interface after hydration. pH dependent contact angle confirmed an existence of carboxylic acid groups in the surface region. The receding angle transition appearing in the pH range of 4-6 was a consequence of hydrophilicity change due to interconverting from carboxylic acid (-COOH) to carboxylate ion (-COO⁻). The surface compositions of imidazole-labeled polymers as analyzed by XPS were consistent with the bulk stoichiometry of the polymers. Reactivity of acid groups towards chemical reaction at the surface was also investigated. The acid groups at the surface of polymers were capable of adsorbing a significant amount of calcium ion from simulated body fluid and being activated by a reaction with *N*-hydroxysuccinimide.

Keywords: biodegradable; carboxylic acid; surface characterization; tyrosine; wettability

Introduction

Desaminotyrosyl-tyrosine ethyl ester (DTE) has been used as a non-toxic, biocompatible monomer for the preparation of tyrosine-derived polycarbonates such as poly(DTE carbonate).[1] Due to its favorable mechanical and thermal properties and processability, poly(DTE carbonate) can be used as a material for orthopedic implants. Due to its slow rate of degradation, poly(DTE carbonate) is not suitable for some biomedical applications such as

DOI: 10.1002/masy.200451210

drug delivery, tissue regeneration scaffold.[2] To overcome this limitation, a new series of tyrosine-derived polycarbonates has been developed based on the rationale that the degradation can be improved by incorporating carboxylic acids as pendent groups.[3] The increased hydrophilicity of the polymer is believed to accelerate the polymer degradation via hydrolysis. The carboxylic acid groups can also serve as versatile precursors for a wide range of chemical modifications, including the attachment of amino acids or peptides. This new series of polycarbonates containing both desaminotyrosyl-tyrosine ethyl ester (DTE) and desaminotyrosyl-tyrosine (DT) is referred to as poly(DTE-co-x%DT carbonate) where x is the mole percentage of the comonomer having a free acid pendent chain.

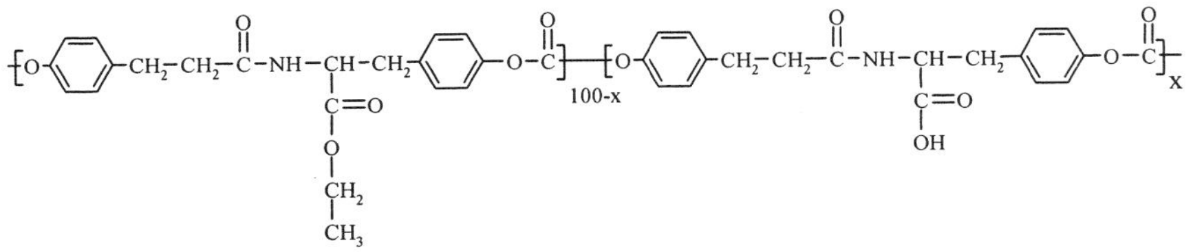

Scheme 1. Structure of poly(DTE-co-x%DT carbonate).

In order to understand the biological properties of a material, it is necessary to know the surface properties since biological systems such as proteins and cells interact first with the surface of the material. In general, a polymer containing two or more functionalities with distinctly different surface free energy in the repeat unit has a strong tendency to exhibit a specific surface composition that may not be consistent with its bulk composition. The fact that poly(DTE-co-DT carbonate) consists of both hydrophobic ester groups and hydrophilic carboxylic acid groups leads to the possibility of surface reconstruction under specific conditions. Such behavior may have significant influences on surface reactivity as well as biological responses.

We have previously reported on the surface characterization of a series of tyrosine-derived polycarbonates containing ethyl, butyl, hexyl and octyl esters as pendent groups.[4] As indicated by contact angle analysis, the degree of surface hydrophobicity was related to the length of the alkyl pendent chain. Elemental composition at the surface as determined by XPS was consistent with stoichiometry of the polymer in the bulk. In this report, the wetting behavior of poly(DTE-co-DT carbonate) is determined by contact angle measurements. The amount of carboxylic acid is qualitatively identified by surface titration and ATR-IR. The surface composition of polymer is characterized by XPS analysis of imidazole-labeled polymer surfaces. The reactivity of carboxylic acid in the surface region is assessed by the determination of adsorbed calcium ions from simulated body fluid and the reaction with N-hydroxysuccinimide.

Experimental

Materials and Methods

All chemicals were purchased from Aldrich and used as received. Films of poly(DTE-co-DT carbonate) (Mw ~80,000-100,000) were prepared by solution casting from 10% (w/v) polymer in methanol/methylene chloride (1/10 v/v). Contact angle measurements were made with a Rame'-Hart telescopic goniometer and a 24-gauge flat-tipped needle. Water purified using a Millipore Milli-Q system and buffer solutions (Metrepak, Micro Essential Laboratory) were used as probe fluids. Dynamic advancing (θ_A) and receding (θ_R) angles were recorded while the probe fluid was added and withdrawn from the drop, respectively. X-ray photoelectron spectra (XPS) were obtained with a Kratos XSAM-8000 instrument with a monochromatic MgKα radiation. Spectra were taken at a takeoff angle of 45° between the plane of the sample surface and the entrance lens of the detector optics. Attenuated total reflectance-Fourier transform infrared spectroscopy (ATR-FTIR) was performed using a Mattson Cygnus 100 Spectrometer equipped with ZnSe crystal. All spectra were obtained from 200 scans at 2 cm^{-1} resolution. The spectra were corrected for atmospheric water and carbon dioxide absorption, then resolved into gaussian components by means of an iterative least-squares method (GRAMS 386 software; Galactic Industries)

Labeling of Carboxylic Acid Groups *via* Formation of Carbonyl Imidazolide

To a nitrogen-purged Schlenk flask containing polymer films, a solution of 1,1'-carbonyldiimidazole (CDI) (0.15 g, 0.93 mmol) in 30 mL THF/diethyl ether (1/4 v/v) was introduced. The reaction proceeded at ambient temperature. After 8h, the films were rinsed sequentially with three aliquots of THF/diethyl ether (1/4 v/v) and hexane and dried under vacuum for 24h.

Adsorption of Ionic Species from Simulated Body Fluid

A polymer film was incubated in a small glass vial containing 10 mL freshly prepared simulated body fluid (SBF) from 0.071 g Na_2SO_4, 0.174 g K_2HPO_4, 0.353 g $NaHCO_3$, 0.368 g $CaCl_2$, 0.305 g $MgCl_2.6H_2O$, 7.99 g NaCl, 0.224 g KCl, 6.06 g Tris and 45 mL HCl (1.0N) at 37°C. After the desired incubation time, the film was removed and rinsed thoroughly by two aliquots of SBF followed by three aliquots of water. The film was dried under vacuum for 24h.

Reaction with *N*-hydroxysuccinimide

To a nitrogen-purged Schlenk flask containing polymer films, a solution of *N*-hydroxysuccinimide (NHS) (0.0955 g, 0.5 mmol) and 1-ethyl-3-(3-dimethylaminopropyl)-carbodiimide (EDCI) (0.0575 g, 0.5 mmol) in 10 mL ethanol was added. The reaction proceeded at ambient temperature to yield *N*-hydroxysuccinimide ester on the surface. After the desired reaction time, the films were rinsed thoroughly with ethanol and dried under vacuum for 24h.

Results and Discussion

Advancing/receding water contact angles (θ_A/θ_R) as a function of acid content are shown in Figure 1. The advancing angle of polymers containing up to 50%DT was not significantly changed as %DT increased. The angles were only varied in the range of 77±2°. Even though θ_A was reduced to ~71° for poly(DT carbonate)(100%DT), the value was not as low as what can be expected from the surface bearing such hydrophilic acid groups. This data implied that the hydrophobic parts of the polymer (main chain & side chain) are the dominating species at the polymer/air interface especially when %DT is below 50%. The carboxylic acid groups preferably reside in the subsurface in order to minimize the surface free energy. It is also possible that the density of carboxylic acid group is quite low considering a relatively large overall dimension of the polymer repeat unit so that they do not contribute significantly to the wetting of polymer. Upon hydration, the carboxylic acid side groups can expose themselves to the polymer/water interfaces such that a significant reduction of the receding angle was observed; 47° for poly(DTE carbonate) to 20° for poly(DT carbonate). Apparently, the receding angle can be a better indication of the change in surface hydrophilicity than the advancing angle as a function of %DT. In general, functional groups at polymer/air and polymer/liquid interfaces arrange themselves in such ways that the surface energy at the interfaces are minimized as long as the rotational energy barrier of functional groups in the surface region can be overcome. The rising of contact angle hysteresis ($\theta_A - \theta_R$) is also evidenced. In this particular study, the heterogeneous nature of the copolymers due to functional group variation is regarded as a major factor causing contact angle hysteresis. We do not take surface roughness into consideration since an unusually abrupt change of contact angle was not observed.

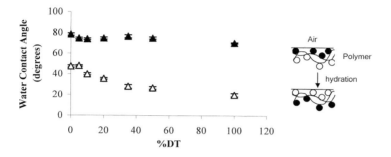

Figure 1. Water contact angles of poly(DTE-co-DT carbonate) as a function of %DT in the bulk: θ_A(▲), θ_R(△). ● and ○ represent ethyl ester side groups and carboxylic acid side groups, respectively.

To verify that an increased hydrophilicity was originated from ionizable carboxylic acid groups in the surface region, pH dependent contact angle analysis or surface titration was performed using buffered solutions. The method has been developed for characterizing the ionizable functional groups.[5] In the case of a surface having carboxylic acid groups, a water contact angle is expected to decrease as a consequence of –COOH being transformed to –COO⁻ after exposure to a basic solution. Without carboxylic acid groups, poly(DTE carbonate) did not exhibit advancing or receding contact angle transition. According to Figure 2, such a transition can be seen on both advancing and receding angle for poly(DT carbonate). Since the acid groups are only accessible at polymer/water interfaces in the cases of polymers containing up to 50%DT, the transition was observed on the receding angles, not on the advancing ones (as shown in Figure 3). It is quite interesting to see that the transition shifts towards lower pH when the acid content is increased.

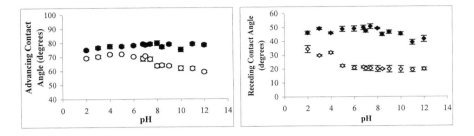

Figure 2. Water contact angles of poly(DTE carbonate) : $\theta_A(\bullet)$, $\theta_R(\circ)$ and poly(DT carbonate): $\theta_A(\bullet)$, $\theta_R(\diamond)$ as a function of pH.

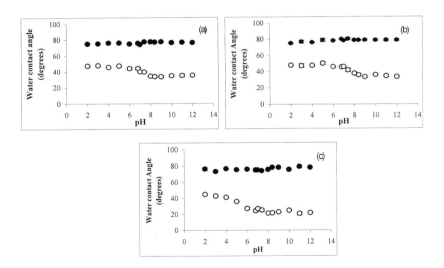

Figure 3. Water contact angles of poly(DTE-co-x%DT carbonate) as a function of pH when (a) x = 10, (b) x = 20 and (c) x = 50 : $\theta_A(\bullet)$, $\theta_R(\circ)$.

According to the ATR-IR spectra shown in Figure 4, the O-H stretching of hydrogen-bonded carboxylic groups which appeared as a broad baseline in the region of 3600-2200 cm^{-1} was seen for polymers containing acid side groups. The magnitude of peak broadening corresponded very well with the acid content suggesting that the carboxylic acid functionality can be qualitatively identified by ATR-FTIR analysis.

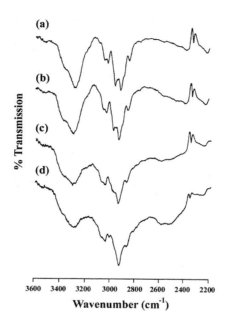

Figure 4. ATR-FTIR spectra of poly(DTE-co-x%DT carbonate) : (a) x = 0, (b) x = 20, (c) x = 50 and (d) x = 100.

The labeling reaction (Scheme 2) was performed by reacting polymer films with CDI in THF/diethyl ether (1:4 v/v).[6] The reaction introduced two additional nitrogen atoms into each DT unit that can be quantified by XPS. The percentage of DT unit in the surface region is equal to (14(%N)-50)/1-(%N/100). % DT tabulated from XPS data was not distinguishable from the bulk composition (analyzed by ^{1}H NMR) for polymers containing 20 and 35% DT, while a slightly underestimated value of 44% DT was detected for poly(DTE-co-50%DT carbonate).

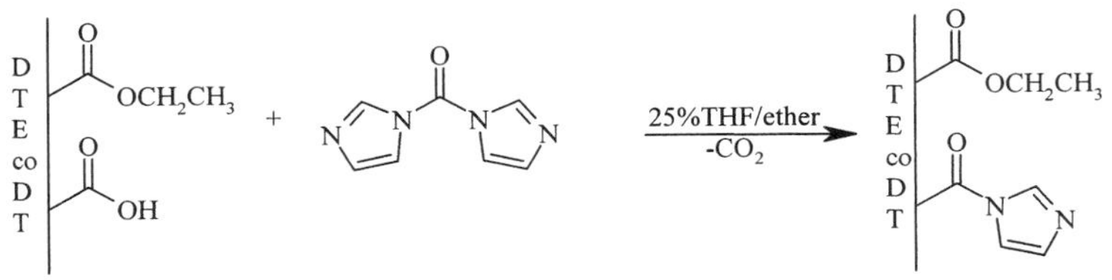

Scheme 2. Labelling reaction of poly(DTE-co-x%DT carbonate) with 1,1'-carbonyldiimidazole.

The reactivity of surface carboxylic acid groups in the form of carboxylate ions (pKa < 6 from surface titration) in physiological pH (~7.4) was assessed by XPS analysis of the polymer films after soaking in SBF. According to Table 1, as high as 0.67% of calcium was detected on poly(DTE-co-50%DT carbonate) after 3-day incubation. The XPS survey spectrum is shown in Figure 5. It should be noted that 0.9% of calcium should be found if there is one calcium ion (Ca^{2+}) binding ionically with every two acid groups. Interestingly, there were no signals from other cationic species such as Na^+, K^+, Mg^{2+}. Due to the limited swelling of poly(DTE-co-20%DT carbonate) in aqueous solution of SBF, 7 days of incubation was required to reach the adsorption equilibrium at ~0.2%. The fact that no calcium signal was observed on poly(DTE carbonate) implied that non-specific physical adsorption did not occur. Due to its rapid degradation, poly(DTE-co-50%DT carbonate) became mechanically unstable after 7-day incubation.

Table 1. Percentage of Ca_{2p} from XPS analysis of polymers after incubation in SBF.

Polymer	Percentage of Ca_{2p} after incubation in SBF		
	1 day	3 days	7 days
Poly(DTE carbonate)	0.00	0.02	0.02
Poly(DTE-co-20%DT carbonate)	0.03	0.05	0.21
Poly(DTE-co-50%DT carbonate)	0.44	0.67	-

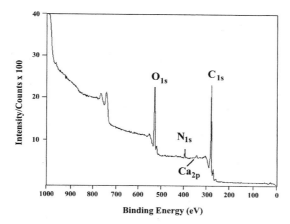

Figure 5. XPS spectrum of poly(DTE-co-50%DT carbonate) after incubation in SBF at 37°C for 3 days.

A coupling reaction with NHS has been recognized as one of effective methods used for converting the carboxylic acid groups into active esters before reacting with amino-containing compounds (Scheme 3). Poly(DTE-co-20%DT carbonate) was chosen as a representative for reactivity assessment. To increase the swelling of poly(DTE-co-20%DT carbonate), the polymer was exposed to ethanol. This pretreatment allowed the reaction to proceed into sufficient depth so that the success of the surface activation could be assessed by ^{1}H-NMR.

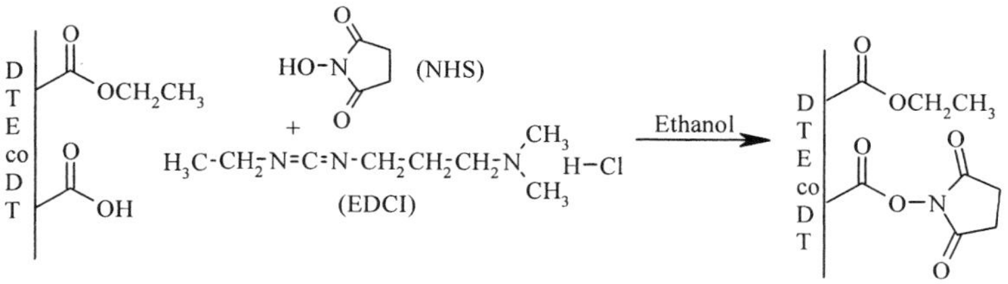

Scheme 3. Reaction of poly(DTE-co-x%DT carbonate) with NHS/EDCI.

The extent of reaction was determined from relative peak area at 2.85 ppm corresponding to the protons of N-succinimide and those binding to the α-carbon of tyrosine at 4.8 ppm. Data displayed in Figure 6 suggested that the higher the concentration of NHS and EDCI, the higher the extent of the reaction can be achieved. The yields of reaction were 0.5, 2.8 and 8.6% when 10, 20 and 50 mmol (in 10mL ethanol) of NHS and EDCI were used, respectively. It is worth noting that the calculated reaction yields based on ^{1}H NMR data represent the information from the bulk so reaction yields at the surface should be considerably higher.

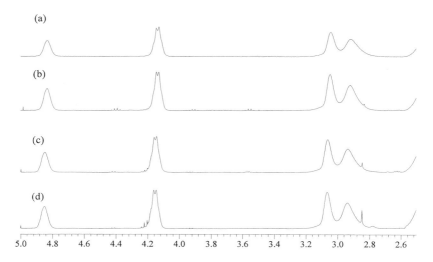

Figure 6. ^{1}H NMR spectra of poly(DTE-co-20%DT carbonate) after reaction with NHS/EDCI: (a) 0 mmol, (b) 10 mmol, (c) 20 mmol and (d) 50 mmol.

Conclusion

It has been proven that surface hydrophilicity of the tyrosine-derived polycarbonate can be enhanced by an introduction of carboxylic acid as pendant groups. The presence of acid component in the surface region can be effectively identified by pH-dependent contact angle measurements, ATR-FTIR and XPS analyses. The ability to adsorb calcium ions in proportion to the acid content suggested that these acid-containing polymers are very useful for orthopedic applications whose healing process basically involves hydroxyappatite formation. According to the reactivity assessment by a reaction with N-hydroxysuccinimide, it can be concluded that the carboxylic acid groups in the surface region are reactive enough for further chemical modification.

Acknowledgements

The initial phase of this research work was supported by a grant from the National Institutes of Standards and Technology (NIST) to Integra Lifesciences Corporation (Plainsboro, New Jersey). The authors also acknowledge support from the New Jersey Center for Biomaterials (http://www.njbiomaterials.org) and Rachadapisek Sompoj Endowment Fund from Chulalongkorn University for Organic Synthesis Research Unit.

[1]   S. I. Ertel, J. Kohn, *J. Biomed Mater Res.* **1994**, *28*, 919.
[2]   V. Tangpasuthadol, S. Pendharkar, R. Peterson, J. Kohn, *Biomaterials* **2000**, *21*, 2379.
[3]   J. Kohn; D. Bolikal, Unpublished data.
[4]   V. H. Perez-Luna, K. A. Hooper, J. Kohn, B. D. Ratner, *J. Appl. Polym. Sci.* **1997**, *63*, 1467.
[5]   S. R. Holmes-Farley, R. H. Reamey, T. J. McCarthy, J. Deutch, G. M. Whitesides, *Langmuir* **1985**, *1*, 725.
[6]   J. –M. Levӓsalmi, T. J. McCarthy, *Macromolecules* **1997**, *30*, 1752.

Macromol. Symp. **2004**, *216*, 99-107

Surface Modification of Chitosan Films-Grafting Ethylene Glycol Oligomer and Its Effect on Protein Adsorption

Wimonsiri Amornchai,[1] *Vipavee P. Hoven,*[2] *Varawut Tangpasuthadol*[*2]

[1]Petrochemistry and Polymer Science Program, Faculty of Science, Chulalongkorn University, Bangkok 10330, Thailand
[2]Organic Synthesis Research Unit, Department of Chemistry, Faculty of Science, Chulalongkorn University, Bangkok 10330, Thailand
E-mail: varawut.t@chula.ac.th

Summary: This work is a part of a series on surface modification of materials made of chitosan. This report focused on grafting monomethoxy ethylene glycol oligomers (mPEG) on the surface of chitosan films. The chemical reactions were performed by immersing the films in organic solvent containing aldehyde derivative of mPEG. The presence of ethylene glycol moieties was determined by attenuated total reflectance-infrared spectroscopy (ATR-IR) and nuclear magnetic resonance (NMR). The hydrophobicity of the modified surface, determined by air-water contact angle, decreased when the ethylene glycol derivatives were grafted on the film. The modified films were also subjected to protein adsorption study in order to assess their uses in biomedical applications. It was found that the presence of ethylene glycol units reduced the adsorption of proteins (albumin and lysozyme) on the films. We therefore have shown that manipulation of the interaction between chitosan and bio-macromolecules is possible by chemically modifying the surface of chitosan.

Keywords: biomaterials; chitosan; poly(ethylene glycol); protein adsorption; surface modification

Introduction

Surface modification of biomaterials has been a main interest for many years, since it is the surface of these materials that first comes into contact with the biological surrounding. A number of techniques have been used to alter the chemical composition and thus, the surface property of the materials. Some of the methods include plasma treatment,[1] blending with other macromolecules,[2,3] and immobilizing small or large molecules on the surface.[4-6] The change of surface property was found to affect adsorption of platelets[2] cells[7-9] and biomacromolecules, such as, proteins,[9-12] on the polymer surface. Chitosan is naturally originated and characterized as non-toxic biomaterials with good biocompatibility. The

DOI: 10.1002/masy.200451211

structure of chitosan contains a large number of hydroxy and amino groups , it can thus be modified by various chemical reactions. It can be hypothesized that the reactions of chitosan films with certain molecules having different chain length and flexibility are able to modify the surface properties of the films, and hence, affect the interaction between chitosan and bio-macromolecules. We have reported earlier[13] methods for grafting hydrocarbon molecules onto the chitosan film. Stearoyl chloride, succinic anhydride, and phthalic anhydride were covalently linked to the amino groups of chitosan, i.e. by amide bond (See Scheme 1). It was found that the type of molecules formed on the surface could somewhat affected the amount of BSA (bovine serum albumin) and lysozyme adsorbed on the films.

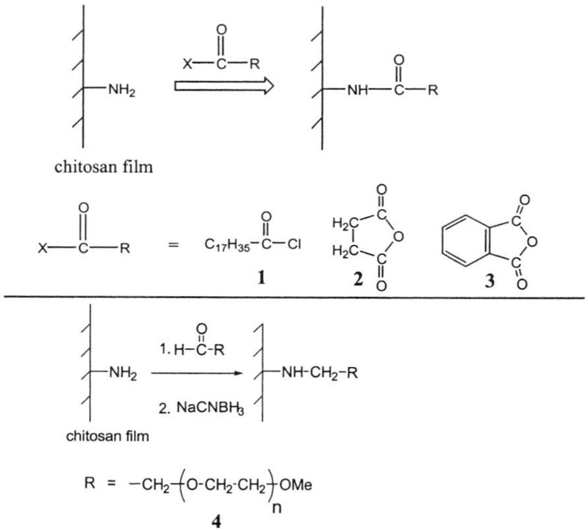

Scheme 1. Surface modification of chitosan films by the reaction of chitosan and various compounds.

Poly(ethylene glycol) (PEG) has been well recognized as one of a polymeric component that can effectively suppress non-specific protein adsorption. The ability to repel protein is believed to originate from the large excluded volume of the hydrated PEG and its chain mobility. In order to extent the use of chitosan in the biomedical applications in which non-fouling is critically required, monomethoxy poly(ethylene glycol)s [$CH_3O(CH_2CH_2O)_nH$ or mPEG] in the form of aldehyde derivatives were grafted onto the chitosan films. Ethylene glycol oligomers having the molecular weight of 550 (n = 12) and 164 (n = 3) were used in this study. The films before and after modification were characterized by ATR-IR and NMR for functional group analysis, and by water contact angle measurement for determining surface hydrophilicity. Protein adsorption study was also performed in order to investigate the reduction of protein fouling.

Experimental

Materials

Oxalyl chloride, sodium cyanoborohydride were purchased from Fluka Chemika. Bovine serum albumin, lysozyme, bicinchoninic assay kit, phosphate buffer saline (PBS), and triethylamine (TEA) were purchased from Aldrich Chemical Co. Monomethoxy poly(ethylene glycol) (mPEG) (M_w = 550) and monomethoxy triethylene glycol (mTEG) (M_w = 164) were obtained from Fluka Chemika. Chitosan (M_v = 645,000; 87%DD) was purchased from Seafresh Chitosan (Lab) Co., Ltd., Thailand.

Preparation of Chitosan Films

Chitosan (2 g) was dissolved in 0.1 M acetic acid (100 mL). After stirring for 24 h, the solution was filtered through a medium pore size sintered glass to remove insoluble substances. The solution was then poured into a Teflon-coated mold. Solvent was allowed to evaporate in air for 4 days. Then the film was washed with 0.1 M NaOH/methanol (1:1) and methanol/water (1:1) to neutralize the acid. The final drying step was carried out under vacuum. Film thickness was between 40 to 100 μm.

Synthesis of mTEG-ald

To the solution of oxalyl chloride (35 mmol) in CH_2Cl_2, DMSO (70 mmol) in CH_2Cl_2 were carefully added under N_2 and cooled in a dry-ice-acetone bath. The solution was stirred for 5 min. Then a solution of mTEG (29.7 mmol) in CH_2Cl_2 was added dropwise. The mixture was stirred for 3 h. TEA (144 mmol) was added dropwise over a period of 20 min. The reaction mixture was left for 30 min at −78 °C and then allowed to reach room temperature. The crude product was concentrated by rotary evaporator to become viscous and colorless liquid. yield: 60% ^{1}H NMR δ 3.1 (3H, s, OCH_3), 3.3-3.4 (8H, t, OCH_2CH_2), 3.9 (2H, s, C$\underline{H}_2$CHO), 9.6 (1H, s, CHO).

Synthesis of mPEG-ald

The same oxidation procedure for mPEG was similar to the one used for synthesizing mTEG-ald. Viscous and colorless liquid was obtained after the crude mixture was concentrated by a rotary evaporator. yield: 50% ^{1}H NMR δ 2.9 (3H, s, OCH$_3$), 3.1-3.3 (40H, m, OCH$_2$CH$_2$), 3.8 (2H, s, C$\underline{H}_2$CHO), 9.6 (1H, s, CHO).

Grafting of mTEG and mPEG on Chitosan Films (4)

A solution of mTEG-aldehyde or mPEG aldehyde (aldehyde eq. = 10 folds of –NH$_2$ units of chitosan) was dissolved in methanol and was added into a flask containing chitosan films. The mixture was stirred for 30 min at rt. Sodium cyanoborohydride (10 eq.) in 5 mL methanol was added to the reaction mixture dropwise for 20 min. The solution was stirred for 3 more days at rt. Chitosan films were later washed in methanol, and dried in vacuo. The second set of experiment was carried out by increasing the ratio of glucosamine unit:-aldehyde:NaCNBH$_3$ from 1:10:10 to 1:30:30.

ATR-FTIR Analysis

All IR spectra were collected at a resolution of 4 cm^{-1} and 16-scan using Bruker Vector 33 FT-IR spectrometer equipped with a DTGS detector. A multiple attenuated total reflection (MATR) accessory with 45° zinc selenide (ZnSe) IRE (Spectra Tech, USA) and a variable angle reflection accessory (Seagull^{TM}, Harrick Scientific, USA) with a hemispherical ZnSe IRE were employed for all ATR spectral acquisitions.

Air-Water Contact Angle Measurement

A goniometer model CAM-PLUS MICRO was used to measure air-water contact angle in a static mode. All measurements were performed at 22-25 ^{o}C. The reported data were averages of 8 measurements.

Protein Adsorption Study

Film substrates were pre-immersed in PBS solution for 2 h. The adsorption experiments were performed in polyethylene bottles, by immersing the film in the protein solution and incubated at 37 °C. After 3 h, the films were removed and rinsed with 10 mL buffer 4 times. Each film was then immersed in 1.0 wt% sodium dodecyl sulfate (SDS) for 1 h at rt, followed by sonication for 10 min to remove the adsorbed protein. Micro-bicinchoninic acid (BCA) protein assay was utilized to determine the amount of protein, using a UV-spectrometer with microtiter plate reader (model Sunrise; Tecan Austria GmbH) at a wavelength of 562 nm.[14]

Results and Discussion

Synthesis of Ethylene Glycol Aldehyde Derivatives

DMSO-oxalyl chloride (called *Swern Oxidation*) is the most widely used DMSO-based reagents for the oxidation of primary and secondary alcohols to aldehydes and ketone, respectively. It usually gives excellent yields with short times and minimal formation of byproduct. The conversion of the terminal hydroxyl group to an aldehyde was determined by the presence of aldehyde proton signal at 9.6 ppm using ^{1}H-NMR spectroscopy. The yields were found to be 60% for mTEG and 50/% for mPEG. The presence of trace amount of water in the ethylene glycol oligomers is believed to retard the oxidation. It should be noted here that the oxidized product could not be removed from the reaction mixture. Therefore, both ethylene glycol reactant and the aldehyde product were present during the grafting step.

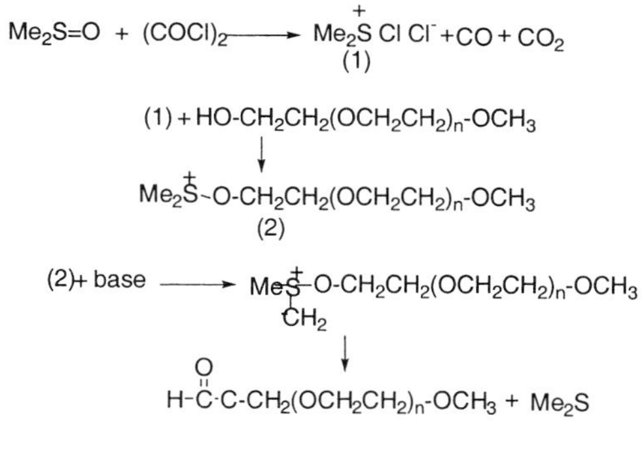

$$Me_2S{=}O + (COCl)_2 \longrightarrow Me_2\overset{+}{S}\,Cl\,Cl^- + CO + CO_2$$
$$(1)$$

$$(1) + HO{-}CH_2CH_2(OCH_2CH_2)_n{-}OCH_3$$
$$\downarrow$$
$$Me_2\overset{+}{S}{-}O{-}CH_2CH_2(OCH_2CH_2)_n{-}OCH_3$$
$$(2)$$

$$(2) + base \longrightarrow Me\overset{+}{S}{-}O{-}CH_2CH_2(OCH_2CH_2)_n{-}OCH_3$$
$$\overset{|}{C}H_2$$
$$\downarrow$$
$$H{-}\overset{O}{\overset{\|}{C}}{-}C{-}CH_2(OCH_2CH_2)_n{-}OCH_3 + Me_2S$$

Scheme 2. Swern oxidation of mTEG to mTEG-aldehyde.

Grafting of Ethylene Glycol on the Chitosan Films

The reductive alkylation of chitosan with mTEG-ald or mPEG-ald in the presence of NaCNBH$_3$, a reducing agent, is shown in Scheme 3.

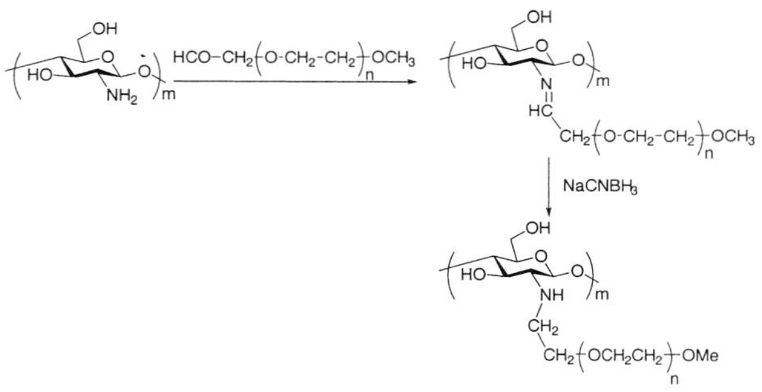

Scheme 3. Grafting of monomethoxy ethylene glycol aldehyde on chitosan.

ATR-IR was used to characterize the functional groups on the surface of chitosan films (with the sampling depth of ~ 1 μm) before and after modification. In Fig. 1A, the non-modified chitosan film showed signals at 1650 and 1590 cm^{-1} for the C-O stretching (amide) and N-H bending (amine) respectively. Analysis of the modified films (Fig. 1B and 1C) revealed a slightly more intense signal of C-H deformation at 1450 cm^{-1} than the non-modified sample. The slight increase of the 1450 signal could be due to the presence of ethylene glycol segments attached to the film surface. However, the differences between the non-modified and modified films are not conclusive. This could be due to the fact that the chemical structure of chitosan is very similar to the ethylene glycol units.

In order to obtain more evidences of the grafting, we analyzed the modified chitosan by ^{1}H-NMR (Fig. 2). The signals at 2.4 and 3.1 ppm indicated $-CH_2-$ unit of ethylene glycol that was immediately next to $-NH-$ and $-NH_2^+-$ of chitosan, respectively. An appearance of these signals suggests that the modification proceeds to a significant depth from the surface so that the attachment can be demonstrated by NMR, which is a bulk characterization technique.

Surface Hydrophilicity

Air–water contact angle measurements were used to determine the hydrophilicity of the film surface. In general, the water contact angle of a hydrophobic surface is higher than that of the hydrophilic surface. From Table 1, the hydrophilicity increased as expected after the chitosan films reacted with the hydrophilic mTEG-ald and mPEG-ald. It seems that increasing the molecular weight and mole equivalent of ethylene glycol units do not affect the hydrophilicity of the films.

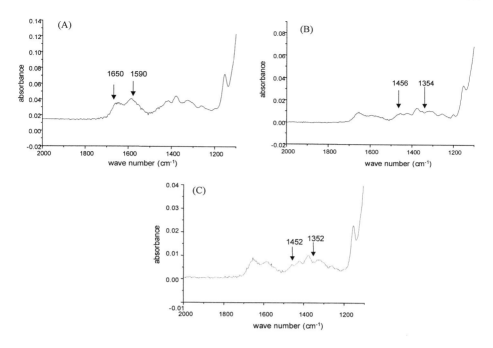

Figure 1. ATR-IR spectra of (A) unmodified chitosan film, (B) mTEG-grafted chitosan film, and (C) mPEG-grafted chitosan film.

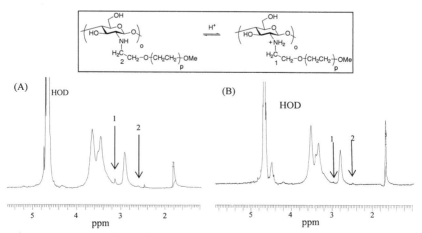

Figure 2. ^{1}H NMR spectra of modified chitosan films after grafting with mPEG-ald (A) and mTEG-ald (B) (solvent: 1% CD_3COOD in D_2O, 25 ^{o}C).

Table 1. Air-water contact angle of chitosan films and modified chitosan films (8 repetitions).

Film samples	Solvent	Contact angle (degree)
Chitosan	-	78.4 ±3.2
Chitosan:mTEG-ald:NaCNBH$_3$		
1:10:10	MeOH	58.4 ±2.9
1:30:30	MeOH	64.1 ± 2.6
Chitosan:mPEG-ald:NaCNBH$_3$		
1:10:10	MeOH	59.3 ± 2.5
1:30:30	MeOH	60.7 ± 1.4

Protein Adsorption Study

Albumin and lysozyme were the two proteins chosen for adsorption study on the chitosan films. Both proteins are model globular proteins that vary in size and charge as well as conformational stability under the experimental condition (pH 7.4 buffer). Bicinchoninic acid assay was used to measure the amount of protein.

In Table 2, it was found that PEG molecule was able to reduce the amount of protein adsorbed on the films. This finding agreed well with previous reports,[15, 16] which indicated that the flexible ethylene glycol units could hinder the approaching protein molecules.

Table 2. The amount of proteins adsorbed on the unmodified and modified chitosan films.

Film samples	Amount of Albumin (μg/cm^2)			Amount of Lysozyme (μg/cm^2)		
	Set I	*Set II*	Average	*Set I*	*Set II*	Average
Chitosan	*1.60*	*1.53*	1.57	*2.81*	*2.19*	2.86
Chitosan-g-mTEG	*0.36*	*0.42*	0.39	*1.28*	*0.34*	0.81
Chitosan-g-mPEG	*0.84*	*0.71*	0.78	*2.34*	*1.65*	2.00

In conjunction with our previous study on surface-modified chitosan,[13] we have found that choices of chemical compounds attached to chitosan could affect its response to protein adsorption as follow

- Hydrophobic (N-stearoyl chitosan) surface enhances protein adsorption.
- Imide bond on chitosan surface could be hydrolyzed to produce carboxylic groups which could interact differently to proteins having different isoelectric points.
- Ethylene glycol oligomer units grafted on chitosan surface could reduce protein adsorption on the film.

Conclusion

In this study, chitosan films were modified by immersing the solvent-cast films in a solution containing aldehyde derivatives of monomethoxy ethylene glycol oligomers. The presence of ethylene glycol on the films were monitored by ATR-IR and ^{1}H-NMR. The PEG-grafted films became more hydrophilic than the original chitosan. Protein fouling on the film also decreased when ethylene glycol oligomers were attached. We have proved here that the surface of chitosan materials can be modified by attaching a certain chemical group in order to manipulate its response to bio-macromolecules, such as proteins that are commonly found in blood of human and animals.

Acknowledgments

The authors would like to thank Asst. Prof. S. Ekgasit for ATR-IR analysis, Asst. Prof. T. Vilaivan for his helpful comments and suggestions on the synthesis, and Department of Material Science for the use of contact angle apparatus. Funding was provided by Radjadapiseksompoj Research Fund and Research Fund from the Graduate School, Chulalongkorn University.

[1]   H. Wang, Y.-E. Fang, Y. Yan, *J. Mat. Chem.* **2001**, *11*, 1374.
[2]   D. Anderson, T. Nguyen, P.-K. Lai, M. Amiji, *J. Appl. Polym. Sci.* **2001**, *80*, 1274.
[3]   É. Kiss, I. Bertóti, E. I. Vargha-Butler, *J. Colloid Interface Sci.* **2002**, *245*, 91.
[4]   K. Hojo, M. Maeda, Y. Mu, H. Kamada, Y. Tsutsumi, Y. Nishiyama, T. Yoshikawa, K. Kurita, L. H. Block, T. Mayumi, K. Kawasaki, *J. Pharm. Pharmacol.* **2000**, *52*, 67.
[5]   X. Qu, A. Wirsén, B. Olander, A.-C. Albertsson, *Polym. Bull.* **2001**, *46*, 223.
[6]   A. Higuchi, K. Shirano, M. Harashima, B. O. Yoon, M. Hara, M. Hattori, K. Imamura, *Biomaterials* **2002**, *23*, 2659.
[7]   T. Koyano, N. Minoura, M. Nagura, K. Kobayashi, *J. Biomed. Mater. Res.* **1998**, *39*, 468.
[8]   G. Haipeng, Z. Yinghui, L. Jianchun, G. Yandao, Z. Nanming, Z. Xiufang, *J. Biomed. Mater. Res.* **2000**, *52*, 285.
[9]   G. Khang, J.-H. Choee, J. M. Rhee, H. B. Lee, *J. Appl. Polym. Sci.* **2002**, *85*, 1253.
[10] H. Sashiwa, J. M. Thompson, S. K. Das, Y. Shigemasa, S. Tripathy, R. Roy, *Biomacromolecules* **2000**, *1*, 303.
[11] R. G. Chapman, E. Ostuni, M. N. Liang, G. Melulent, E. Kim, L. Yan, G. Pier, H. S. Warren, G. M. Whitesides, *Langmuir* **2001**, *17*, 1225.
[12] G. Ladam, P. Schaaf, F. J. G. Cuisinier, G. Decher, J.-C. Voegel, *Langmuir* **2001**, *17*, 878.
[13] V. Tangpasuthadol, N. Pongchaisirikul, V. P. Hoven, *Carbohydr. Res.* **2003**, *338*, 937.
[14] P. K. Smith, R. I. Krohn, G. T. Hermanson, A. K. Mallia, F. H. Gartner, M. D. Provenzano, E. K. Fujimoto, N. M. Goeke, B. J. Olson, D. C. Klenk, *Anal. Biochem.* **1985**, *150*, 76.
[15] M. M. Amiji, *Carbohydr. Polym.* **1997**, *32*, 193.
[16] S.-i. Aiba, X. Mo, P. Wang, K. Hayashi, Z. Xu In *3rd Asia Pacific Chitin and Chitosan Symposium* Taiwan, **1998**; Vol. 3, p 424-429.

Synthesis and Characterization of a Novel Poly(aryl ether) Containing 4-Chloro-2,5-diphenyloxazole

Nuttaporn Pimpha,[1] *Supawan Tantayanon,*[*1] *Frank Harris*[2]

[1] Functional Polymer and Petrochemistry Laboratory, Department of Chemistry, Faculty of Science, Chulalongkorn University, Bangkok 10330, Thailand
E-mail: supawan.t@chula.ac.th
[2] The Maurice Morton Institute of Polymer Science, Akron University, Ohio, USA

Summary: 4-Chloro-2,5-bis(4-fluorophenyl)oxazole monomer has successfully been synthesized using cyclization reaction of 4-fluorobenzoyl cyanide with 4-fluorobenzaldehyde. This monomer was converted to poly(aryl ether)s by nucleophilic substitution of the fluorine atoms on the benzene rings of oxazole monomer with bisphenol A. The influence of the reaction time on the molecular weight had been investigated. The polymers were identified by FT-IR,^{1}H-NMR and TGA. The products exhibited weight-average molar masses up to 2.81×10^4 g mol^{-1} in GPC. These poly(aryl ether)s showed very high thermal stability up to 363 °C for 5 % weight loss in TGA under N_2.

Keywords: 2,5-diphenyloxazole; poly(aryl ether); poly(aryl ether oxazole)

Introduction

Poly(aryl ether)s are well-known high-performance engineering thermoplastics. These materials have interesting physical properties, including a high modulus, toughness, high thermal and thermooxidative stability, and chemical resistance.[1] It has been shown that aromatic nucleophilic substitution reaction between activated aryl halide monomer and bisphenoxides can lead to the formation of linear poly(aryl ether)s.[2-3] Thus, the halogens must be activated, which is achieved by the presence of an electron withdrawing group in the para position. Heterocyclic rings can serve as activating groups to decrease the electron density in the para position of adjacent phenyl rings.[4-13] A common feature of these groups is stabilization negative charges developed at the 2- or 4-position of aryl moiety in the transition state of the nucleophilic halogen displacement reaction. Among the heterocyclic rings, 2,5-diphenyloxazole is particularly interesting because it is highly fluorescent and, thus, has potential in fluorescent sensors, laser dyes and scintillators for detecting nuclear radiations.[14-20] Because of these interesting properties of 2,5-diphenyloxazole, considerable

DOI: 10.1002/masy.200451212

attention has been devoted to the preparation of a new classes of 2,5-diphenyloxazole containing polymers.

The 2,5-bis(4-fluorophenyl)oxazole monomer is extremely reactive, because the fluorine atom is activated not only by the negative inductive effect but also by the electron deficiency of the oxazole ring and therefore produces high molar mass poly(aryl ether)s when reacted with bisphenol A.[21] Numerous methods for the preparation of oxazole rings with various substitution patterns have been reported. For example, the cycloaddition reaction of benzonitriles and benzoyl chloride led to the formation of the polymer. However, such synthesis involve several steps and the difficultly to handle CH_2N_2. In this research, the much simple method has been used to synthesize 4-chloro-2,5-bis(4-fluorophenyl)oxazole. We chose to synthesize our monomers according to a modification of Fischer synthesis[22] from the 4-fluorobenzoyl chloride and 4-fluorobenzaldehyde in one step. The objective of this work was the synthesis of poly(aryl ether) containing 4-chloro-2,5-bis(4-fluorophenyl)oxazole was to be prepare and polymerized with bisphenol A.

Experimental

Materials

4-Fluorobenzaldehyde, 1,3-dimethyl-3,4,5-tetrahydro-2(1H)-pyrimidinone (DMPU) and 4-fluorobenzoyl chloride (Aldrich Chemical Co.) were used as received. Potassium carbonate and cuprous cyanide (Fischer Chemical Co.) were ground and dried at 100°C under reduced pressure overnight before use. Bisphenol A (Aldrich Chemical Co.) was recrystallized from methanol. Toluene (Fischer Chemical Co.) was dried with magnesium sulfate overnight before use. Tetrahydrofuran was freshly distilled from sodium/benzophenone.

Instrumentation

Thermogravimetric analyses (TGA) were carried out under nitrogen at a heating rate of 10°C/min using a TA instrument model 2950 thermogravimetric analyzer. Gel permeation chromatography (GPC) was carried out on a Jasco 880-PU system with Jasco UV-970 Detector. Calibration was done with standard polystyrene samples. Tetrahydrofuran was used as the eluant at a flow rate of 1 mL min^{-1} at 28°C. ^{1}H NMR and ^{13}C NMR spectra were recorded on a Brucker AC200 spectrometer. Deuterated chloroform was used as the solvent, and chemical shift values (δ) were reported in parts per million relative to the residual signals of this solvent (δ 7.24 for ^{1}H and δ 77.0 for ^{13}C). Infrared spectra were recorded on an Impact 410 Nicolet FT-IR spectrometer as a solid suspended in a potassium bromide disk. Melting points were measured using an Electrothermal 9100 melting point apparatus.

Synthesis of 4-Chloro-2,5-bis(4-fluorophenyl)oxazole

4-Fluorobenzoyl cyanide (25 mmol, 3.72g) and 4-fluorobenzaldehyde (20 mmol, 3.10 g) in 50 mL dry THF were contained in a reaction flask which was connected to a HCl gas generator. The reaction mixture was allowed to cool in an ice-NaCl cooling bath to 0°C and then dry HCl gas was passed into the solution until the HCl gas was no longer absorbed by the reaction mixture(3 hours), tested by ammonia at the outlet of the reaction flask. At this stage, the reaction flask was quickly sealed and kept at 0°C for 2 days. The reaction mixture was then poured onto an ice with continuous stirring and subsequently filtered. The product was recrystallized from 1:1 mixture of hexane/ethylacetate to afford white crystals.(2.36 g, 44%) mp 110°C. MS m/z: 292. IR (KBr): ν (cm^{-1}): 3035, 1621, 1530, 1494, 1439, 1399, 1337, 1238, 1120, 1056, 905, 850, 644 and 524. ^{1}H NMR (300 MHz, CDCl$_3$): δ (ppm) 8.09 (4H, m) and 7.13 (4H, m). ^{13}C NMR (200 MHz, CDCl$_3$): δ (ppm) 164.4, 163.1, 160.8, 150.9, 138.5, 132.1, 128.7, 126.4, 124.8, 124.3 and 116.4.

Polymerization

Oxazole monomer (0.73 g, 2.5 mmol) and bisphenol A (0.57 g, 2.5 mmol) were dissolved in 20 mL of DMPU and 10 mL of toluene under nitrogen in a round-bottom flask with a reflux condenser, a nitrogen inlet, and a Dean-Stark trap. Potassium carbonate was added (0.75 g, 5.1 mmol), and the mixture was refluxed for 3 h at 180 °C in an oil bath temperature. During refluxing, fresh and dry toluene was refilled twice. Finally, the temperature was raised to 220 °C while the toluene was removed through the Dean-Stark trap. After 12 h, 1 mL of acetic acid was added to neutralize the phenoxide, and the mixture was filtered hot to remove salts. After the reaction was cooled to ambient temperature the polymer was precipitated by pouring the mixture into a mixture of 450 mL of methanol and 50 mL of water. The polymer was filtered off, washed with water and methanol and dried under reduced pressure. IR (KBr): ν (cm^{-1}): 3052, 1608, 1512, 1486, 1473, 1333, 1250, 1207, 1174, 1132, 1053 and 824. ^{1}H NMR (300 MHz, CDCl$_3$): δ (ppm) 8.05(m), 7.65(m), 7.25(m), 7.05(m), 6.95(m) and 1.71(s).

Results and Discussion

Synthesis of monomer

To synthesize 4-chloro-2,5-bis(4-fluorophenyl)oxazole, 4-fluorobenzoyl cyanide has firstly been obtained from nucleophilic substitution reaction between 4-fluorobenzoyl chloride and cuprous cyanide.[23] It was then reacted with 4-fluorobenzaldehyde as shown in Scheme 1.

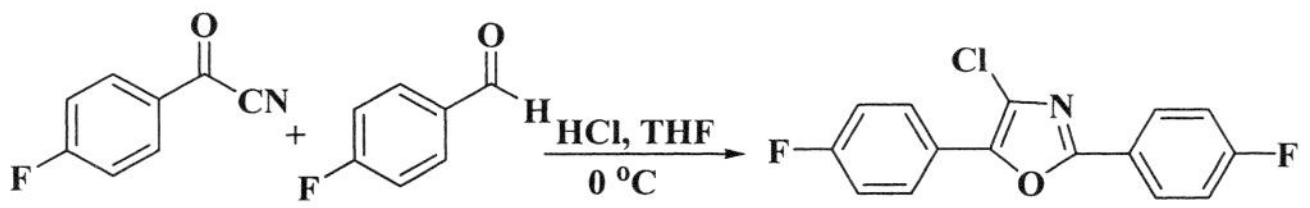

Scheme 1. Reaction scheme and structure of 4-chloro-2,5-bis(4-fluorophenyl)oxazole monomer.

The IR spectrum of this product confirmed the existence of an oxazole ring as shown by a strong absorption band at 1530 cm^{-1} assigned to the –N=C-O- ring stretching frequency. The mass spectrum of this compound contained m/z peaks corresponding to the calculated molecular weights of 292 and 294 with the ratio of 3: 1 due to the presence of chlorine in the molecule. ^{1}H and ^{13}C NMR data also indicated the structure of 4-chloro-2,5-bis(4-fluorophenyl)oxazole. In particular, it was clearly revealed by the ^{13}C signals at 160.8 (C2), 150.9 (C4) and 138.5 (C5) ppm which are the characteristic ^{13}C pattern for 4-chloro-2,5-diphenyloxazole compounds synthesized in our laboratory. The ring formation has been proposed to occur via the formation of acylimidoyl chloride and then reacted further with the carbonyl carbon of the aldehyde. Apparently, 4-chloro-2,5-bis(4-fluorophenyl)oxazole can successfully be synthesized in one step with a simple work-up. It should be noted that the fluorine of 4-fluorobenzaldehyde served as the electron donating group in this reaction which retarded the ring formation. That's the reason why, the reaction afforded only 44% yield. This explanation can be supported by another reactions between benzoyl cyanide and several other 4-substituted benzaldehydes. For examples, the reaction of benzoyl cyanide with 4-dimethylaminobenzaldehyde and terephthalaldehyde gave 20% and 70% yield, respectively. The synthesis of the 4-chloro-2,5-diphenyloxazole compounds and the mechanism of their formation will be reported in a separate paper.[24]

Synthesis of polymer

Polymerization of the 4-chloro-2,5-bis(4-fluorophenyl)oxazole with stoichiometric amounts of bisphenol A was carried out in the presence of potassium carbonate in DMPU/toluene as shown in Scheme 2.

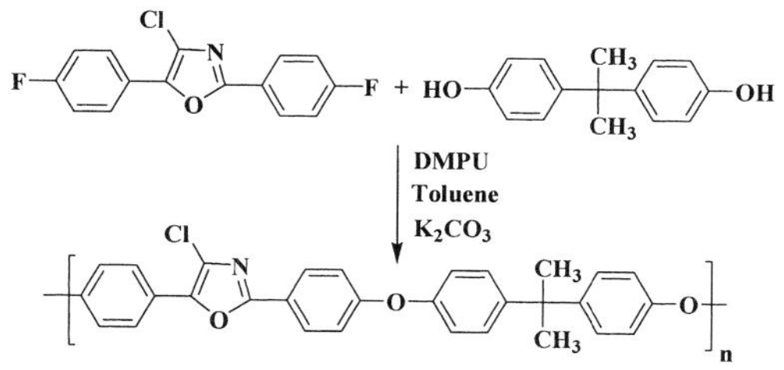

Scheme 2. Reaction scheme of the poly(aryl ether) containing 4-chloro-2,5-bis(4-fluorophenyl)oxazole unit.

During the initial stage of the polymerization, the reaction temperature was maintained at 180-200°C to convert bisphenol A to its salt. The reaction was drived by the removal of water from yhe reaction mixture as the azeotropic mixture with toluene via a Dean-Strak trap. Upon completion of the salt formation and dehydration, the reaction temperature was raised to 220°C to effect the nucleophilic displacement. The influence of polymerization time on molecular weight of the polymer was observed. At lower reaction time, the lower molecular weightof the polymer was obtained as shown in Table 1.

Table 1. Polymerization results at different reaction time.

Reaction time (h.)	Mw (g/mol)	TGA (5% weight loss)
3	17,844	341
6	13,295	334
9	28,118	363
12	13,115	359

This was presumably due to the lack of sufficient time necessary to maximize the chain growth. At polymerization time greater than 9 h, however, the decrease in molecular weight was resulted. It was probably caused by three possible chain scissions, the attack of KF and the carbonate from base at the polymer chains and transetherification. Similar results on condensation polymerization of the other aromatic difluoride with phenoxides has been reported.[25]

Thermal stability

The poly(aryl ether)s exhibit excellent thermal stability as summarized in Table 1. These polymers showed a 5% weight loss at the temperature ranging from 334 to 363°C as determined by dynamic TGA. Maier et al. reported that similar poly(aryl ether oxazole) without a chlorine group on an oxazole unit exhibited decomposition temperatures at 466 °C in nitrogen atmosphere.[8] The source of the instability of the oxazole ring in the polymers can be found by studying the synthetic routes to oxazoles.

Cycloaddition reactions are known to be reversible at high temperatures and therefore oxazoles may be subject to this kind of decomposition reaction to benzonitrile and ketonecarbene. In the case of P_1 and P_2, one of the two most important mesomeric structures of the ketocarbene is strongly stabilized by chlorine. Therefore, the activation energy for the cycloreversion of the chlorinated oxazole is lower than that for the cycloreversion of the unsubstituted oxazole, and hence the decomposition of the chlorinated oxazole starts at lower temperatures.

ketonecarbene

Solubility

All poly(aryl ether)s containing 4-chloro-2,5-diphenyloxazole show good solubility in a number of common organic solvents. Table 2 summarizes the results of qualitative solubility tests which were attempted to obtain 10% w/v solutions of the polymers in various solvents. They were easily soluble in a number of common organic solvents including THF and chloroform, as well as in more polar amidic solvent such as NMP.

Table 2. Solubility tests with the Poly(aryl ether oxazole)s.

Reaction time	NMP	DMSO	THF	Acetone	CHCl$_3$	Toluene
3	+	+	+	+	o	o
6	+	+	+	+	o	o
9	+	+	+	+	o	o
12	+	+	+	+	o	o

(+) soluble at room temperature
(o) soluble on heating

Conclusion

In this paper, we reported here a new 4-chloro-2,5-bis(4-fluorophenyl)oxazole monomer which has been synthesized using cyclization reaction between 4-fluorobenzoyl chloride and 4-fluorobenzaldehyde. Polymer synthesis was successful and the 4-chloro-1,3-oxazole can activate nucleophilic displacement in the synthesis of poly(aryl ether oxazole)s. The optimum polymerization time to obtain a high molecular weight up to 2.63 x 10^5g mol^{-1} was 9 hour. A 5 % weight loss occurred between 334-363°C in TGA under N_2.

Acknowledgments

The authors of this paper would like to thank the Thailand Research Fund under the Golden Jubilee number PHD/225/2541 for financial support of this research.

[1] Susanta, B.; Maier, G. *Chem Mater.***1999**, 11, 2179.
[2] Susanta, B.; Maier, G.; Martin, B. *Macromolecules*. **1999**, 32, 4279.
[3] Maiti, S.; Mandal, B. *Prog. Polym. Sci.* **1986**, 12,111.
[4] Hedrick, J.L.; Twieg, R. *Macromolecules*. **1992**, 25, 2021.
[5] Hedrick, J.L.; Labadie, J.W. *Macromolecules*. **1988**, 21, 1883.
[6] Arnold, F.E.; Van Deusen, R.L. *Macromolecules*. **1969**, 2, 497.
[7] Blaise, C.; Bouanane, A.; Brembilla, A.; Lochon, P.; Neel, J. *J. Polym. Sci., Polym. Symp.* **1975**, 52, 137.
[8] Maier, G.; Schneider, J. M. *Macromolecules* **1998**, 31, 1798.
[9] Hedrick, J. L. *Macromolecules.***1991**, 24, 6361.
[10] Saegusa, Y.; Iwasaki, T.; Nakamura, S. *J.Polym. Sci., Part A: Polym. Chem.* **1994**, 32, 249.
[11] Hedrick, J. L.; Labadie, J. W. *Macromolecules.***1990**, 23, 2223.
[12] Hedrick, J. L.; Jonsson, H.; Karter, K. R. *Macromolecules.* **1995**, 28, 4342.
[13] Hilborn, J. G.; Labadie, J. W.; Hedrick, J. L. *Macromolecules.* **1990**, 23, 2854.
[14] Clapman, B.; Sutherland, A. J. *Tetrahedron Lett.* **2000**, 41, 2257.
[15] Clapman, B.; Sutherland, A. J. *Tetrahedron Lett.* **2000**, 41, 2253.
[16] Hiromitsu,I.; Tomiki,I.; Kunihiro,I. *Macromolecules.* **1993**, 26, 4533.
[17] Clapman, B.; Richards, A. J.; Wood, M. L.; Sutherland, A. J. *Tetrahedron Lett.* **1997**, 38, 9061.
[18] Horrocks, D. L. *Applications of Liquid Scintillation Counting*, Academic Press, New York, **1974**, 51.
[19] McCairn, M. C.; Hine, A. V.; Sutherland, A. J. *J. Mater. Chem.* **2003**, 13, 225.
[20] Hamerton, I.; Hay, J. N.; Jones, J. R.; Lu, S. *J. Mater. Chem.* **1998**, 8, 2429.
[21] Maier, G.; Hecht, R.; Nuyken, O.; Helmereich, B.; Burger, K. *Macromolecules* **1993**, 26, 2583.
[22] Micheal, D.; Ram,L.; Bela,T. *J.Heterocyclic Chem.*, **1977**, 14,317.
[23] Oalewood, T. S.; Weizgerber, C. A. *Organic Syntheses*. Coll 3, 113.
[24] Pimpha, N.; Tantayanon, S. *Heterocycles*. Accepted for Publication.
[25] Klein, D. J.; Modarelli, D. A.; Harris, F.W. *Macromolecules* **2001**, 34, 2427.

Micromechanical Analysis of Thermoplastic - Thermoset Interphase

J. M. Vazquez-Rodríguez,* P. J. Herrera-Franco, P. I. Gonzalez-Chi

Centro de Investigación Científica de Yucatán, AC, Unidad de Materiales, Calle 43, No. 130, col. Chuburná de Hidalgo 97200, Mérida, Yucatán, México
E-mail: ivan@cicy.mx

Summary: The interfacial shear strength of a pull out model between a thermoplastic fiber and thermoset matrix was analyzed. The method to analyze the interfacial quality in this kind of composite was Photoelasticity. The interfacial shear strength was measured localizing the isocromatic fringes. The Isochromatic fringe corresponds to the points along the specimen in which the principal stresses have the same value.

Keywords: composites; interfacial shear strength; photoelasticity; polyesters; pull out

Introduction

Photoelasticity is a well-established technique for stress analysis and has a wide range of industrial and research applications. The main goal of this technique is the determination of isoclinic and isochromatic fringes to analyze the shear stress and stress distribution in a sample [1, 2]. The light used for the analysis can be white or a bean with only some monochromatic lines. In the first case, the analysis generates an image with a full decomposition of wavelengths in which, each one corresponds to a level of shear strength. It is easer to recognize the phase change for the image analysis of the shear strength when a narrow-chromatic light is used [3].

Photoelasticity has been used to study the shrinkage process of resins used for medical items. To understand this phenomenon, could improve composite elaboration for clinical effectiveness [4, 5].

A photoelastic experimental method has been used for analyzing the stress at the vicinity of crack tips and, in conjunction with the theory of Shear Lag, the stress fields between a short-fiber and the matrix in composites were described [6, 7, 8, 9]. Photoelasticity is a common way

© 2004 International Union of Pure and Applied Chemistry

DOI: 10.1002/masy.200451213

for testing a single fiber composite. The analysis gives the opportunity to follow *in-situ* their failure mechanics like fiber debonding, fibre breaks, matrix fracture and matrix yield [10].

Wang Feng and Xie Himin [11] made a strain analysis for materials with different elastic modulus using the method of the shear stress difference to calculate the interfacial strength. The model was based on the experimental solution of the stress tensor obtained from an interphase deformed continuous conditions, reducing the tensor to a plane problem, by vertical projection of one of the stress components. For the analysis models for the axial and transversal direction were needed [11].

Photoelasticity was also used for monitoring the failure process due the humidity in carbon fiber and epoxy resin composites [12]. Heterogeneous fibre composites were studied and it was found that the specimen had an equivalent birefringence and an equivalent birefringence direction to the addition of the different fibre distribution [13, 14].

Some other studies used an interferometric technique for photoelastic testing to detect the spatial deformation of the stressed specimen and hence the spatial strain distributions. This study concluded that the maximum stressed zones are those were the maximum fringe distortion occurs [15, 16]. A. Tomlinson and E. A. Paterson [17] used a different method to analyze the shear strength. They used a step-wise loading system and the model was sectioned into discrete imaginary-slices and a relationship was derived between the retardation and the isoclinic angle [17].

There had been attempts to digitalize and process the images with software to obtain reconstructed theoretical fringe patterns from a loading frame in a polarization bench [18, 19]. Nevertheless it requires frame recorders and some other equipment that makes the process complicated.

Polarization

A plane beam is obtained using a polariscope, which consists of a light source and two polarization elements with the polarization axis crossed. A single light beam composed of a large number of randomly oriented waves, propagates following a harmonic motion along a straight line [20, 21]. Polarization fixes the light beam propagation in only single plane. The photoelastic patterns can be obtained with a plane or a circular polariscope, and the light intensity (I) emerging is:

$$I = E^2 sen^2 2\alpha \ sen^2 \frac{R}{2} \qquad (1)$$

The equation (1) represents the dark field of a plane polariscope in which I is function of the angle between the normal and the principal stress direction (α) and the relative retardation (R) induced when the light beam passes through the stressed specimen. There are two conditions in which I is zero (the light beam is extinguished). The first one is when the angle of the principal stress directions corresponds to α_i= 0, $\pi/2,\pi$, $3\pi/2$ . . . radians and the second condition is when the relative retardation corresponds to R= 0, 2π . . . $N\pi$ radians, where N is an integer. N represents the number of complete wavelengths or retardation cycles produced by an induced strain and it is called isochromatic fringe.

When the polariscope axis coincides with the principal stress direction (α) a dark fringe is obtained. This fringe corresponds to the points along which the principal stresses have parallel directions and it is known as isoclinic fringe [21].

In the case of the isochromatic order N, when a complete cycle of retardation is reached, it produces another dark fringe called an isochromatic fringe which is directly related to the shear strength. N is known as photoelastic order, fringe order or isochromatic order. The isochromatic fringes are easier to obtain using a circular polarization, which is obtained by adding two-quarter wave elements to a plane polariscope. The light intensity I emerging from this polariscope is:

$$I = E_y^2 (1 - \cos R)$$
$$I = 2E_y^2 sen^2 \frac{R}{2}$$
$$I = K sen^2 \frac{R}{2} \qquad (2)$$

Where K is a constant for each polariscope. This equation has the following extinction conditions for the fringe order N:

$$I = 0 \qquad for \qquad \frac{R}{2} = 0, \pi ...radians \quad or \quad R = 0, 2\pi ...., N(2\pi)$$

When the relative strain induces a retardation of N_i= 1, 2, 3 . . . a dark isochromatic fringe is obtained. The stress optic law [20] is commonly written as:

$$\sigma_1 - \sigma_2 = \frac{Nf_\sigma}{h_L} \qquad (3)$$

where $N = R/2\pi = \delta/\lambda$ is the relative retardation in terms of complete cycles N and depends from the wavelength (λ). h_L is the distance traveled by the light in the stressed specimen. $f_\sigma = \lambda/C$ is the material fringe value or fringe-stress coefficient. And C is the ratio between the relative stress-optic coefficients of each material.

The shear strength is defined by the following equation:

$$\frac{\sigma_1 - \sigma_2}{2} = \frac{N f_\sigma}{2 h_L} = \tau_{Max} \quad (4)$$

Photoelastic Calibration

The photoelastic calibration of the epoxy resin was made using the four points bending method. This method produces stress patterns of pure bending: The beam has no vertical stress (σ_y), transverse normal stress (σ_x) or horizontal shear stresses (τ_{xy}).

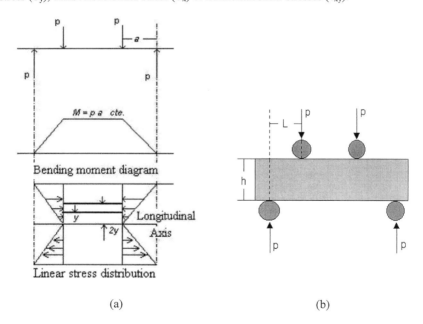

(a) (b)

Figure 1. Bending moment diagram, linear stress distribution (a), of four points-bending set up (b).

The equation used for the epoxy resin calibration was:

$$f_\sigma = \frac{3Pa}{h^2 N} \quad (5)$$

Where P is the load in Newton, a is the length between supports in meter, h is the beam height in meter, N is the photo elastic order and f_σ is the fringe-stress coefficient in Newton per meters.

Pull-out Test

The micromechanic technique used was the Pull-out test in which one end of a filament is embebed in a matrix block and the filament free end is loaded. The applied load and the displacement are continuously monitored until the filament is extracted or the interphase fails. The stress analysis is made considering a balance between the tension stress at the fiber (σ_f) and the shear stress (τ) at the interphase [3]. The Pull-out test has been used to measure the adhesion between two different materials[22, 23] and it is widely used to measure the shear strength at the interphase between a fibre and an epoxy-matrix[24, 25, 26].

Materials

The Pull-out specimens were manufactured using a bisphenol-A epoxy resin DER 331 from DOW Química Mexicana S. A. The curing agent was an aliphatic diamine ANCAMINE 1784 from Air Products and Chemicals, Inc. The thermoplastic polyester fiber was from KIRSCHBAUM.

Experimental

The fiber tensile test was performed according to the ASTM D2343-67 using a universal test machine Shimatzu A 61 fitted with a 5 kN load cell at 30 mm/min with a gauge length of 245 mm.

The tensile test of the epoxy resin was made using the ASTM D 638-82[a] in the universal test machine Shimatzu A 61 with a load cell of 5 kN, gauge length of 50 mm. The testing speed was 1 mm/min. The Poisson ratio was measured according to the ASTM D 638-82[a] with two extensometric gauges (EA-06250 BF-350 gauge factor 2.095 ± 0.5%) cemented perpendicularly to each one along the main specimen axis. The test was in a step-wise loading fashion at the elastic zone of the epoxy resin.

To calculate the Poisson ratio the equation used was:

$$v_{yx} = -\frac{\varepsilon_x}{\varepsilon_y} \quad (6)$$

Where v_{xy} is the Poisson ratio, ε_x and ε_y are the strains in the axial and transversal directions respectively.

Polariscope Arrangement

The polarization instrument was arranged as follows to conform a circular polariscope. Two polarization elements crossed 90°. Two quarter-wave elements displaced ± 45° to the polarization axis of the first polarization element. The light emission was obtained form a mercury bulb with a wavelength of 540 nm. The load was transmitted to the specimen using a load frame with a cantilever system.

Specimen Preparation

The epoxy resin specimen for the optical calibration was 1.0 cm wide, 10.0 cm long and 0.5 cm thick. The pull-out specimen had the following dimensions: The resin block was 4 cm wide, 6 cm long and 1.1 cm thick. The fiber was longitudinally embebed at the center of the block, and the embebed length was of 4 cm. The free length of the fiber was 3 cm. The curing process took place for eight days under controlled humidity conditions at room temperature.

Results and Discussion

The tensile properties of the fiber are shown at Table 1 and for the epoxy resin al Table 2.

Table 1. Tensile characteristics of the polyester fiber.

	Elastic module	**Strength**	**Max. Strain**	**Failure stress**
	MPa	MPa	%	MPa
Average	6003.61	576.779	46.6248	568.637
SD	1.7078	1.7078	1.7078	1.7078

Table 2. Tensile properties of the Epoxy resin.

	Elastic Module	Strength	Max Strain	Poisson ratio
Specimen	MPa	MPa	%	
Average	1032.050	22.7621	9.3126	0.3840
SD	2.5819	2.5819	2.5819	1.8708

The ratio between the elastic module of the fiber and the epoxy resin is around 6, which is enough to get the isochromatic response from the specimens maintaining the elastic behavior of the fiber.

The epoxy resin had an average Poisson ratio of 0.3840. The performance of the epoxy resin block with this Poisson ratio generates a radial compression stress on the fiber during the load process. In the present analysis, the compression stress was not measured and considered low. To analyze the performance of the compression stress in a Pull-out specimen and to understand how affects the failure process it is necessary to analyze the directions of the principal stress.

Photoelastic Calibration of the Epoxy-resin.

Figure 2 shows a picture sequence from the four points bending test. They show the increase of the shear strength at the specimen as an increase on isochromatic fringes.

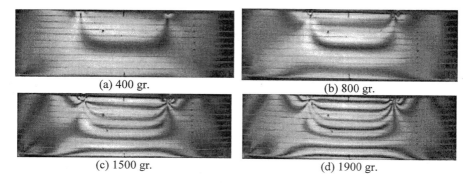

(a) 400 gr.

(b) 800 gr.

(c) 1500 gr.

(d) 1900 gr.

Figure 2. Photoelastic fringes of a circular polariscope (dark plane) in a four points bending system. The applied load is below each picture. The photoelastic order (a) zero, (b) one, (c) two and (c) three.

From the pictures the photoelastic fringe stress-coefficient f_σ was obtained. The dark fringe at the center is the zero photoelastic order; the upper photoelastic orders were counted from the center to the sides.

Figure 3 shows the plot of $3Pa$ vs. h^2N. The slope is the f_σ. coefficient according the equation 5. Five specimens were tested and f_σ was calculated (Table 3), its average value is $f_\sigma = 2,432.62$ N/m (SD = 65.89).

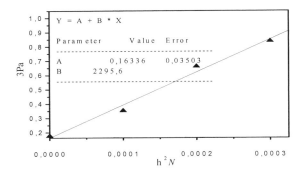

Figure 3. Equation 5 plot to calculate the fringe-stress coefficient f_σ which is the slope of the lineal regression.

Table 3. Fringe-stress coefficient f_σ for specimen 1.

Specimen 1	h = 1 cm		a = 1.5cm		
Weight grams	Load P Newton	Photoelastic Order (N)	3Pa	h^2N	Fringe stress Coefficient f_σ. (slope)
400	3.92	0	0.1766	0	
800	7.85	1	0.3532	0.0001	
1500	14.72	2	0.6622	0.0002	2295.6
1900	18.64	3	0.8388	0.0003	

Shear Strength by Photoelasticity

The strain transmitted from the fiber to the resin block gave us the opportunity to monitor the interphase failure process, which can be divided in three principal steps. The initial step did no show any photoelastic response. An incipient response was obtained when the minimum load used was 400 gr. The following step was the shear strength growth. During this step before of the interphase failure two photoelastic fringes were obtained.

The pictures used to measure the shear strength between the fiber and the epoxy resin blocks are shown at the Figure 4.

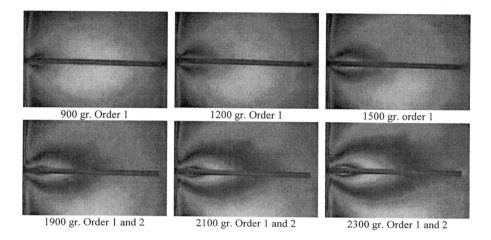

Figure 4. Isochromatic fringes for different loads (dark plane of a circular polariscope).

These pictures are from the loaded Pull out specimens. The beginning and the end of the photoelastic fringe mark the field where the shear strength acts over the fibre. The shear strength is related to the fringe order and it is maximum at the left side at the block where the fiber emerges from the epoxy resin. For that reason the second order fringe (small fringe close to the fiber) appears on that region. The maximum shear strength begins at this area where the free fiber was loaded. From 900 gr. to 1500 gr. the development of the first photoelastic order (N_i=1) is shown. When the load reaches 1900 gr., the second photoelastic order (N_i=2) appears. The second photoelastic order is directly related to higher shear strength, and acts over the same points of the epoxy resin block where the first photoelastic order was before. The growing of the second photoelastic order conduced to the interphase failure; only two photoelastic orders were obtained. The maximum load that the specimens supported before the interphase failure was of 2400 gr. The shear stress for each specimen was calculated using the equation 4.

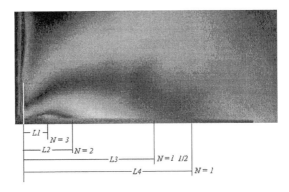

Figure 5. Measuring and location of photoelastic fringes.

Five specimens were tested to measure the photoelastic shear strength. The loads used were from 400 gr. to 2400 gr. To measure the shear strength rise, two photoelastic orders from the dark field and three from the bright field were obtained. The images of the dark field were obtained directly from the polariscope, and the images form the bright field were the negative image of the dark field. The dark fringes in these images represent the half-photoelastic orders of the bright field. The half orders obtained were: ½, 1 ½ and 2½ orders. These five orders (Table 4) were used to find the τ_{Max} distribution along the fiber as is shown in the Figure 5.

Table 4. Shear stress for the specimen 1.

| Specimen 1 | | Fringe stress | |
| h_L = 5.5 mm | | Coefficient f_σ. = 2427.900 N/m | |
Photoelastic order N	$(\sigma_1 - \sigma_2) = 2\,\tau_{Max}$ Strength (Pa)	$\tau_{Max} = N f_\sigma./2h_L$ Shear strength (Pa)	Longitud on the fiber affected by the τ_{Max} L(cm)
0.50	220718.182	110359.091	3.7053
1.00	441436.364	220718.182	2.9413
1.50	662154.545	331077.273	1.4576
2.00	882872.727	441436.364	1.0152
2.50	1103590.909	551795.455	0.6595

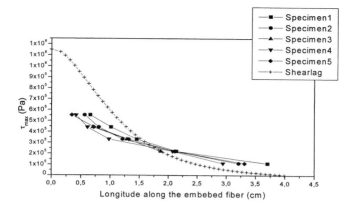

Figure 6. Shear strength distribution along the embebed fibre of the Pull-outspecimen analyzed by photoelasticity.

Figure 6 shows the shear strength distribution from the Shear lag theory and experimental results. The length along which the shear strength acts was measured from the beginning of the photoelastic order at the left side of the pictures (fiber section) to its end in the epoxy resin block. τ_{Max} acting over the section from 0.25 to 0.70 cm from the border over the embebed fiber for photoelastic test. The maximum Photoelastic shear strength is located about 0.5 cm from the border of the specimen. After this maximum, the shear strength decreases towards the embebed end of the fiber.

The performance of the photoelastic Shear strength versus Shear lag follows almost the same trajectory except close to the free fiber section and at the embebed tip where the Shear lag theory do not takes in count variables like the adhesion at the end of the embebed fiber or the complex loads close to the free fiber section.

Conclusions

The photoelastic technique relates the strains with the interfacial shear strength in a Pull out specimen under a load. The isocromatic fringes obtained from a Pull-out specimen were used to measure the interfacial shear strength and in spite of this, we can measure the maximum shear strength for a thermoplastic-thermoset interphase.

The load process of the Pull-out specimens was followed in real time. It made possible to see the initial load and the strength transfer from the fiber to the epoxy resin block. That effect was shown earlier. Where, the initial load was followed for a growing of the strain. The shear strength form the initial strain to the extraction of the embebed-fiber was shown by the isochromatic fringe.

The mechanical characteristics of the embebed fiber and the epoxy resin were different. The elastic module of the fiber was 6003.61 MPa and the module of the epoxy resin was 1032 N/mm^2. This difference helped to strain the epoxy resin and to obtain the photoelastic fringes while the fiber acted as an elastic rod. The Poisson ratio of the epoxy resin was of 0.38403.

The isochromatic fringes for a four point bending system were obtained with a dark field and it was reproducible. The point in which each fringe order was located is important; if the fringe order was located in a different point it conduces to an error of the epoxy-calibration. The photoelastic calibration of the epoxy resin was made and the fringe-stress coefficient f_σ had an average value of 2,427.9 N/m.

There were a good concordance between the Shear lag theory and the photoelastic experiment and the experimental results from the photoelastic experiments were reproducible for all the specimens tested.

[1]   J. A. Quiroga, A. Gonzalez Cano, *Solid Mechanics Solid Mechanics and Its Applications* **2000**, *82*, 17.
[2]   J. A. Quiroga, A. Gonzalez Cano, *Applied Optics* **2000**, *39*, 2931.
[3]   S Yoneyama, M. Shimizu, *Optics and Lasers in Engineering* **1998**, *29*, 423.
[4]   Y. Kinomoto, M. Torij, *Journal of Dentistry* **1998**, *26*, 165.
[5]   Hassan M. Ziada, John F. Orr, *The journal of Prosthetic Dentistry* **1998**, *80*, 661.
[6]   J.-S. Hawong, J.-G. S. Kyungdong, Yeungnam University; Kyungdong Junior College VIII International Congress on Experimental Mechanics Nashville, Tennessee, USA, 10-13 June **1996**
[7]   P. J. Withers, E.M. Chorley, T. W. Clyde, *Materials Science and Engineering A* **1991**, *13* ,5173.
[8]   S. Ritter, Busse, Plenum Publishing Corp., Review of Progress in Quantitative Nondestructive Evaluation. Volume 17B USA **1998**, 1201.
[9]   J. Hobbs, R. Burguete, *Journal of Strain Analysis* **2000**, *36*, 93.
[10]   S. Ritter, Busse Plenum Publishing Corp., Review of Progress in Quantitative Nondestructive Evaluation. Volume 17B USA **1998**, 1201.
[11]   W. Feng, X. Himin, *Journal of Material Processing Technology* **1997**, *65*, 165.
[12]   Z. R. Xu, K. H. G. Ashbee, *Int. Journal of Materials Science* **1994**, *2*, 394.
[13]   W. Zhang, Y. Wan, *Int. Journal of Mechanical Science* **1995**, *37*, 933.
[15]   K. Bhattacharya, A. Basuray, *Optics Communication* **1994**, *109*, 380.
[16]   S. Yoneyama, M. Takashi, *Optics and Lasers in Engineering* **1998**, *30*, 441.
[17]   A. Tomlinson, E. A. Patterson, *Journal of Strain Analysis* **1999**, *34*, 295.
[18]   A. Ajovalasit; S. Barone; G. Petrucci, *Journal of Strain Analysis for Engineering Design*, **1998**, *33*, 75.
[19]   J. A. Quiroga; A. Gonzalez Cano, *Applied Optics* **1997**, *36*, 8397.
[20]   C. P. Burger, A. S. Kobayashi, *"Photoelasticity"* cap 5 Handbook of Experimental Mechanics second revised edition **1993.**
[21]   M. M. Frocht, Ph. D. *"Photoelasticity"*Vol. I John Wiley and Sons, Inc. New York, **1947.**

[22]   P. J. Herrera Franco V. Rao, L. T. Drzal, *Composites* **1992**, *23*, 2.
[23]   M. R Piggott, *Composites Science and technology* **1987**, *30*, 295.
[24]   J. A. Nairn, H. D. Wagner, *Material Science and Engineering Mechanics of Materials* **1997**, *26*, 63.
[25]   L.B Greszczuc,. Composites ASTM STP 452, ASTM **1969**
[26]   P. Laurence, *Journal of Materials Science* **1972**, *7*, 1.

Spectroscopic Study of Di-Imide Hydrogenation of Natural Rubber

Jareerat Samran,[1] *Pranee Phinyocheep,*[*1,2] *Philippe Daniel,*[3] *Daniel Derouet,*[4]
Jean-Yves Buzare[3]

[1] Department of Chemistry, Faculty of Science, Mahidol University, Rama VI Road, Payathai, Bangkok, 10400, Thailand
[2] Institute of Science and Technology for Research and Development, Mahidol University, Salaya Campus, Puthamonthon, Nakornpathom, 73170, Thailand
E-mail: scppo@mahidol.ac.th
[3] Laboratoire de Physique de l'Etat Condensé LPEC (UMR CNRS n° 6087), Université du Maine, Avenue Olivier Messiaen, 72085 Le Mans Cedex 9, France
[4] L.C.O.M. Chimie des Polymères (UMR du CNRS LCO2M n° 6011), Université du Maine, Avenue Olivier Messiaen, 72085 Le Mans Cedex 9, France

Summary: The diimide hydrogenation of natural rubber (NR) was studied by using *p*-toluenesulfonylhydrazide (TSH) as a diimide-releasing agent. The microstructure and the percentage of hydrogenation were studied by Raman, ^{1}H-NMR and ^{13}C-NMR spectroscopic techniques. Quantitative measurements on fraction of hydrogenated part gave the results in good agreement by using these techniques. The results indicated that percent hydrogenation increased with increasing of reaction time and about 80-85 % hydrogenation was achieved when a two-fold excess of TSH was used. The vibrational characteristic of C=C bond of NR is strongly Raman active and noted at 1663 cm^{-1}. The decrease of this signal was clearly observed during the progress of hydrogenation but the vibrational frequency of the *cis* and *trans* structures of the trisubstituted olefin unit of NR can not be differentiated by this technique. While ^{1}H- and ^{13}C-NMR analysis showed that *cis-trans* isomerization of carbon-carbon unsaturation of NR occurred during hydrogenation.

Keywords: diimide hydrogenation; FT-IR; natural rubber; NMR; Raman spectroscopy

Introduction

Chemical modification of unsaturated polymers via hydrogenation is one of the most important methods for altering and optimizing the physical and mechanical properties of the macromolecules. The hydrogenation is also a potential method offering a polymer that can not be prepared by a simple conventional polymerization reaction. An example is the preparation of poly(ethylene-alt-propylene) which can be potentially obtained by fully hydrogenation of 1,4-polyisoprene.[1] Diimide (N_2H_2) is an inorganic reducing agent which has been progressively used for hydrogenation of unsaturated molecules.[2,3] It is considered as a noncatalytic reaction and found to be a convenient hydrogenation method since it can be performed under atmospheric pressure with relatively simple apparatus and procedure. In contrary, catalytic hydrogenation using hydrogen gas in the presence of a noble metal catalyst is rather difficult in handling the reaction, usually involving high pressure and temperature.[1,4]

Natural rubber (NR) which has been known as highly cis-1,4 polyisoprenic structure, has a primary drawback in thermal and oxidative stabilities and poor oxygen and ozone resistance. This is due to the presence of the unsaturation along the molecular chain. Therefore, reduction of the unsaturated units of NR should overcome some of these drawbacks. N.K. Singha et al reported that catalytic hydrogenation of NR using RhCl(PPh3) catalyst increased the thermal stability of the resulting product without affecting its glass transition temperature.[1] Utilization of a diimide hydrogenating agent generated from thermal decomposition of p-toluenesulfonylhydrazide (TSH) for hydrogenation of polybutadiene and polyisoprene has been reported.[2,3] In both cases, an excess amount of TSH is required if complete hydrogenation is expected. The evidence of hydrogenation was examined by IR and NMR spectroscopy.

Generally, spectroscopic techniques i.e. Raman, infrared, ^{1}H- and ^{13}C-NMR can be used for characterization of the microstructure of the chemically modified products.[5,6] Hydrogenation of NR can be therefore extensively investigated by vibrational (i.e. Raman and infrared) and resonance spectroscopic techniques (i.e. ^{1}H- and ^{13}C-NMR) since the characteristic signal of the C=C bond of polyisoprene is very sensitive to its environment.[7,8] However, only a few work has been analyzed by Raman scattering and solid state NMR, including the quantification of the unsaturated units of NR. Therefore, these two techniques have been used for such purposes in this present work and compared to other spectroscopic technique i.e. ^{1}H-NMR in solution.

This article describes the hydrogenation of NR by using diimide as a reducing agent generated from the in situ decomposition of TSH. The progress of reaction was observed by

Raman, FT-IR and NMR both in solution and solid state spectroscopic techniques. The percentage of hydrogenation was determined by Raman, ^{1}H-NMR in solution and ^{13}C-NMR in solid state. The evidence of *cis-trans* isomerization as hydrogenation progress was also investigated by ^{1}H- and ^{13}C-NMR.

Experimental

Hydrogenation

Hydrogenation of NR was carried out as follows; about 1.0 g of rubber was dissolved in 100 ml of xylene (JT Baker). Two folds of *p*-toluenesulfonylhydrazide (TSH, Fluka) as compared to rubber unsaturated units ([TSH]/[C=C] = 2) was added to the solution. The mixture was then stirred and heated to 135°C under nitrogen atmosphere. Samples were taken at various reaction times and precipitated in methanol. The hydrogenated product was purified by dissolving in hexane and reprecipitating in methanol. Finally, the product was dried in vacuum at room temperature before analysis.

Characterization

Raman Spectroscopy

Raman spectra of all samples were recorded with a T64000 Jobin-Yvon multichannel spectrometer adjusted either in simple spectrograph configuration with a 600 lines/mm grating or in triple subtractive configuration for high resolution experiments. Samples were illuminated with a Coherent Argon-Krypton Ion Laser selecting the 647.1 nm lines in order to minimize luminescence contribution to spectra. To improve the signal/noise ratio, each spectrum was accumulated 20 times during 30 sec. The frequency range selected was 500-3200 cm^{-1}. Calibration of the spectrometer was precisely checked on the 520.2 cm^{-1} silicon band and the resolution of the spectra was estimated to be smaller than 1 cm^{-1}. All experiments were performed under microscope using an x 50 long work distance objective (Olympus B x 40 microscope).

FT-IR Spectroscopy

FT-IR spectra were carried out by using Perkin-Elmer system 2000 spectrometer by casting thin film of the sample on NaCl plate. All samples were recorded at 16 scans in the range of 500-4000 cm^{-1} with 4 cm^{-1} spectra resolution in order to obtain a good signal-noise ratio.

Nuclear Magnetic Resonance Spectroscopy in Solution

^{1}H-NMR and ^{13}C-NMR (Bruker DPX-300 NMR spectrometer) spectra were obtained from the samples dissolved in CDCl$_3$ using tetramethylsilane (TMS) as an internal reference.

Nuclear Magnetic Resonance Spectroscopy in Solid State

High resolution solid state experiments were recorded on a Bruker MSL 300 spectrometer operating at 75.47 MHz. The instrument is equipped with a high-power amplifier for proton decoupling. The experiments were carried out using a pulse width of 90° (4 μs) with a repetition time of 4 s. A spectral width of 20 kHz and 16 K data points were used for data collection. The spinning speed of MAS technique was applied at 10 kHz.

Results and Discussion

The diimide molecule (N$_2$H$_2$) is generated *in situ* from the thermal decomposition of *p*-toluenesulfonylhydrazide (TSH) as shown in equation (1) in Figure 1. It can then release a hydrogen molecule directly to the carbon-carbon double bonds of isoprene units as represented in equation (2) in Figure 1.

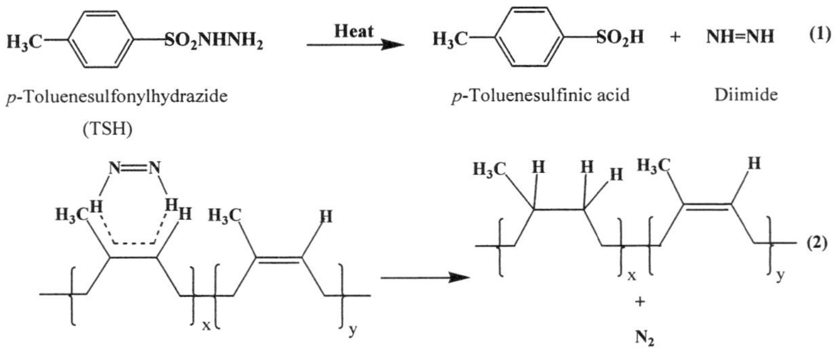

Figure 1. Hydrogenation of NR by using diimide generated from *p*-toluenesulfonylhydrazide.

Microstructure Analysis of Hydrogenated Rubber

Vibrational Spectroscopy

Non catalytic hydrogenation of NR in this study was carried out by using 2 moles of TSH as compared to the isoprene units. When one molecule of diimide reacted with the NR, one unit of the C=C should be disappeared as shown in Figure 1. The vibrational characteristic of the C=C of the isoprene unit can therefore be examined by Raman and Infrared (IR) spectroscopy. A comparison between the Raman spectra of the starting NR and that of hydrogenated NR (HNR) samples taken during hydrogenation at various reaction times is given in Figure 2. As hydrogenation reaction proceeded, the decrease of absorption band at 1664 cm^{-1} assigned to the C=C stretching modes can be clearly detected with increasing of the vibrational intensity of the band at 1432 cm^{-1}, attributed to the -CH$_2$- deformation vibration. No alteration of the band at 1452 cm^{-1} which belongs to an asymmetric vibration of -CH$_3$ group in Raman spectra was noted after hydrogenation.

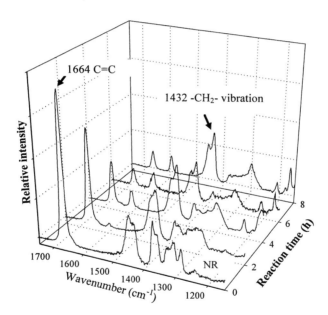

Figure 2. Raman spectra of natural rubber (NR) and hydrogenated rubbers at different reaction times.

It is not surprising that after 8 h of reaction the total disappearance of the peak at 1664 cm^{-1} is not observed as it has been reported that the complete reduction of unsaturation of butadiene polymers and of polyisoprene can be obtained with a five-fold excess of TSH.[2,3]

For IR analysis of the hydrogenated rubber, the decrease of two important characteristic peaks at 1665 and 836 cm^{-1}, attributing to C=C stretching and the C-H out of plane deformation of the trisubstituted olefin of polyisoprene, respectively, were observed as shown in Figure 3. The signal at 1375 cm^{-1} assigning to the C-H deformation and the signal at 735 cm^{-1}, being due to sequences of three continuous methylene units occurred after hydrogenation were also detected. It can be seen that the vibrational absorption modes obtained from IR spectrum is not as powerful as Raman mode. The intensity of the C=C absorption peaks at 1665 and 836 cm^{-1} are not strong for quantitative analysis of the percent of hydrogenation.

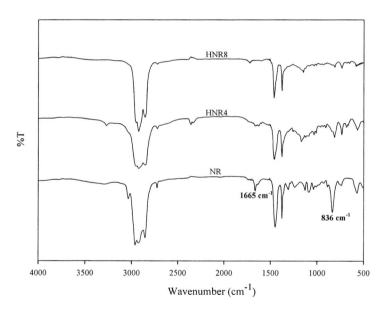

Figure 3. FTIR spectra of natural rubber (NR) and hydrogenated rubbers at 4h (HNR4) and 8h (HNR8) reaction times.

Resonance Spectroscopy

^{1}H-NMR analysis of NR in Figure 4 shows three main characteristic signals of proton adjacent to C=C, methylene and methyl protons of the unsaturated unit at 5.12, 2.03 and 1.67 ppm, respectively. ^{1}H-NMR spectra of the products of hydrogenation after 4 and 8 h of the reaction, symbolized as HNR4 and HNR8, respectively are also shown in Figure 4. The figure indicates that the intensity of proton signal adjacent to C=C bonds at 5.12 ppm decreases with the increased reaction time, as well as the decrease of signal at 1.67 ppm characteristic of methyl proton of *cis*-1,4 polyisoprenic units of NR. The methyl and methylene proton signals observed at 0.84 and 1.1-1.3 ppm, respectively showed a strong increment due to the transformation of double bonds into saturation moieties. The hydrogenation levels can be therefore determined by comparison of the integrals of signal at 5.12 ppm with the integrals for proton signal of saturated units.

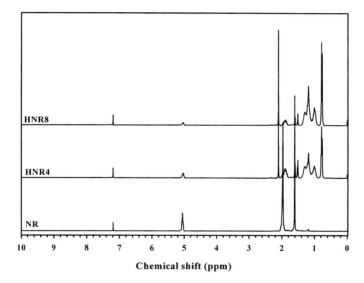

Figure 4. ^{1}H-NMR spectra of natural rubber (NR) and hydrogenated NRs at 4h (HNR4) and 8 h (HNR8) reaction times.

In the case of ^{13}C-NMR analysis in solution, the carbon signals of C=C of *cis*-structure of isoprene units of NR are positioned at 135.2 and 125.5 ppm. Three signals characteristic of methyl and two methylene carbons of the unsaturated unit were found at 23.4, 26.4 and 32.2 ppm, respectively. The extra peaks at 19.5, 33 and 37.1 ppm assigning to the methyl, methine and methylene carbons of saturated units in the hydrogenated rubbers are observed. The detected chemical shifts are in good agreement with those reported in the literatures.[1,9] The ^{13}C-NMR spectrum of the partially hydrogenated product reveals the evidence of the *cis-trans* isomerization of isoprene unit in the polymer chains as the signal at 134.9 and 124.7 ppm, corresponding to olefinic carbons of *trans*-polyisoprenic structure are detected for HNR.

The ^{13}C-NMR study in solid state of natural rubber and hydrogenated rubbers was carried out at room temperature, which is the temperature far above their glass transition temperatures (T_g). Under this condition, the dipolar interactions and chemical shift anisotropy that lead to line broadening are partially averaged by chain motion, and sharp lines are observed with high power decoupling and magic angle spinning. The motional averaging is such that high resolution signals of NR and HNR can be observed with spinning, and peak assignments can be established using the traditional solution methods (Figure 5).

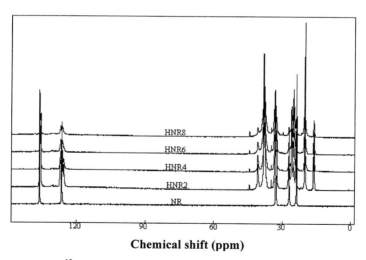

Chemical shift (ppm)

Figure 5. Solid state ^{13}C-NMR spectra of natural rubber (NR) and hydrogenated NR at 2h (HNR2), 4h (HNR4), 6h (HNR6) and 8 h (HNR8) reaction times.

The dominant peaks of NR spectrum at 23.6, 26.7, 32.5, 134.8 and 125 ppm observed were related to the five carbon atoms of the *cis*-1,4 polyisoprenic units, similar to the assignments using ^{13}C-NMR in solution. These characteristic peaks can be seen to diminish as the hydrogenation reaction proceeded. Then the significant peaks at 20.4, 33.3 and 38.1 ppm are found. Based on literature and by making comparison with the NMR spectra in liquid state, these peaks are assigned to the carbon signal of methyl, methine and methylene types, respectively.[1] Figure 5 shows also the characteristic *trans*-1,4-polyisoprene located at 16, 27 and 40 ppm, assigning to methyl and two methylene carbons.[10] This is the confirmation of *cis-trans* isomerization occurred simultaneously with hydrogenation. The extra resonances of trans-structural units allowed us to evaluate the amount of trans-isomer as increasing the reaction time.

Determination of Percent Hydrogenation

With the Raman scattering technique, it is possible to monitor the progress of saturated units in NR during hydrogenation. The percent hydrogenation of each sample can be estimated from the ratio of band areas arising from the stretching mode of C=C bond and the bending mode of CH_2, since each addition of hydrogen molecule on C=C bond gives rise to one new CH_2 unit.[11] The repeating unit of polyisoprene already presents two CH_2 groups. The CH_2 bending band appears to be moderately active near 1432 cm^{-1} as shown in Figure 2. The band intensity in Raman spectra can be roughly considered as proportional to the concentration of each species in the following:

$$A\,(C=C) \quad = \quad k_1[C=C]$$
$$A\,(CH_2) \quad = \quad k_2[CH_2]$$

where A(C=C) is the integrated intensity area of the band located between 1645 and 1685 cm^{-1} assigned to the C=C stretching vibrational band, $A(CH_2)$ is the integrated intensity area of the CH_2 bending band between 1415 and 1440 cm^{-1}, k_1 and k_2 are proportional constants associated with the considered C=C and CH_2 vibrational modes. If the starting rubber is composed of 100 repeating units, 100 C=C and 200 CH_2 units are then present, hence

$$k_1/k_2 \quad = \quad 2A_0(C=C)/A_0(CH_2)$$

Then, the ratio k_1/k_2 can be determined from the Raman spectrum of NR in Figure 2.

X is assigned for the concentration or number of the unit disappeared or formed, when X

C=C units disappear during hydrogenation, it can be noted that X of CH_2 units are created. For a partially hydrogenated product, X or the hydrogenation rate can then be calculated from the following equation :

$$X = 100 \ [(K - 2\alpha)/(K + \alpha)]$$

where $K = k_1/k_2$ and $\alpha = A_0(C=C)/A_0(CH_2)$ of the partially hydrogenated compound.

From ^{1}H-NMR spectroscopy, the integrated areas of the signal at 5.12 and 0.84 ppm, corresponding to the proton adjacent to olefinic unit and methyl proton of saturated unit, respectively, were used for the determination of percentage of hydrogenation. The progress of hydrogenation by solid state ^{13}C-NMR was obtained by using the integrated areas of the signals of saturated methyl carbon at 20.4 ppm comparing to the methyl carbon of both cis- and trans-unsaturated units at 23.6 and 16 ppm, respectively. The results of percentage of hydrogenation at various reaction times of NR determined by different techniques are illustrated in Figure 6. It was found that the hydrogenation increased with the increase of reaction time. The maximal percentage of hydrogenation approximately 80-85 % was found by using all three techniques when a two-fold excess of TSH was used. Several publications reported that complete hydrogenation was obtained when 4-5 moles of TSH per mole of polyisoprene units were utilized for hydrogenation of homopolymer or copolymer containing polyisoprene units.[11,12] It was described that not only the syn form of the generated diimide can react with the C=C bonds of polyisoprene units, but it can also undergo disproportionation, giving nitrogen molecule and hydrazine.

Cis-trans Isomerization

It was reported that utilization of TSH as a diimide releasing agent for hydrogenation of *cis*-polybutadiene resulted in *cis-trans* isomerization.[13] Unfortunately, the *cis-trans* isomerization of the unsaturated units of NR during hydrogenation can not be detected by Raman spectroscopy as the vibrational frequency of the *cis* and *trans* structures of trisubstituted olefinic unit in Raman scattering are very close. While the ^{1}H-NMR spectra of the hydrogenated rubber exhibit the signal of methyl proton of the *cis*- and *trans*-1,4 polyisoprenic units at 1.67 and 1.60 ppm, respectively. The percentage of the remaining double bonds in *cis*- and *trans*- configurations of the hydrogenated samples at various reaction times can then be calculated and the results are shown in Figure 7 (a).

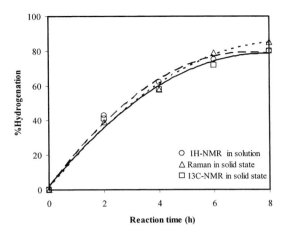

Figure 6. Results of percentage of hydrogenation at various reaction times of NR determined by ¹H-NMR in solution, Raman scattering and ¹³C-NMR in solid state.

The ¹³C-NMR spectrum in solution of the partial hydrogenated product reveals the evidence of the *cis-trans* isomerization. The carbon signals of *cis*-structure were located at 135.2 and 125.5 ppm while the signals at 134.9 and 124.7 ppm, corresponding to olefinic carbons of 1,4-*trans* polyisoprenic units were also detected for hydrogenated NR.

For solid state ¹³C-NMR spectra, the signal of methyl carbon of *cis*- and *trans*- polyisoprenic structure can be clearly seen at 23.6 and 16 ppm, respectively. The results of the percentage of cis- and trans- isomers at various reaction times are illustrated in Figure 7(b). It seems likely that in this system the *cis-trans* isomerization is a reversible process. The thermal decomposition of TSH generates a diimide acted as a hydrogen-donor molecule as well as *p*-toluenesulfonic acid by-product. The formation of unstable complex between the double bonds and the *p*-toluenesulfinic acid by-product may be responsible for the *cis-trans* isomerization reaction.[13] No signal of the addition of the by-product onto the hydrogenated rubber was detected.

142

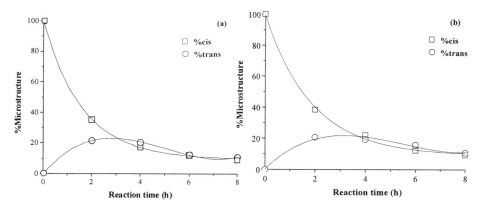

Figure 7. Progress of *cis-trans* isomerization rate during hydrogenation of NR determined by ^{1}H-NMR in solution (a) and ^{13}C-NMR in solid state (b).

Conclusion

Natural rubber (NR) was partially hydrogenated using two-fold moles of *p*-toluenesulfonylhydrazide (TSH) compared to the unsaturated unit. The microstructure of the hydrogenated products can be observed by using spectroscopic techniques i.e. Raman, FT-IR, ^{1}H- and ^{13}C-NMR. The quantitative measurements on fraction of hydrogenated part at different reaction times determined by Raman, ^{1}H- and ^{13}C-NMR gave similar results which indicated that percent hydrogenation increased with increasing of reaction time and about 80-85 % hydrogenation were obtained at 8h. ^{1}H-NMR and ^{13}C-NMR gave the evidence of the *cis-trans* isomerization during hydrogenation. These techniques were also used to monitor the progress of the isomerization.

Acknowledgement

The scholarship from the Ministry of the University Affairs, Thailand to J. Samran is greatly appreciated.

Basic Terms

Retention mechanisms operative in the polymer HPLC can be divided into entropic (exclusion) and enthalpic ones. This classification is highly practical though it says only little about molecular backgrounds of processes taking place within the HPLC column. The **exclusion retention mechanism** includes changes in mixing, conformation [2-5] and possibly also orientation [6] entropy of macromolecules during their passage along the HPLC column. These entropy change result from the concentration gradients, as well as from the flow, and diffusion processes within column and lead to partial or full exclusion of macromolecules from the pores or from the outer surface of packing particles. This **entropic partition** of macromolecules is considered basic retention mechanism of conventional SEC. The **enthalpic retention mechanisms** [7] are connected with the attractive or repulsive energetic interactions among column packing, macromolecules, and eluent molecules. These take place in the mobile phase, and on the packing surface or interface, as well as within the (quasi) liquid stationary phase physically immobilized on or chemically bonded to the packing surface. It is useful to consider a set of binary enthalpic interactions in the system,[7] namely between macromolecules and eluent **(thermodynamic quality of eluent),** between macromolecules and packing **(affinity of analytes toward packing),** and between eluent and packing **(eluent strength).** The thermodynamic quality of a solvent determines solubility of macromolecules and in the first approximation it can be characterized by the exponent in the Kuhn-Mark-Houwink-Sakurada viscosity law, a. The effective segmental interaction energy, ε, describes affinity between macromolecules and column packing. The concept of eluent strength introduced by Snyder is based on the considerations about interaction energy between molecules of solvent and solid surfaces. Interaction of eluent molecules with the bonded stationary phase leads to solvation phenomena. In the case of mixed mobile phases the preferential solvation effect should be considered. Above interactions are responsible for the retention mechanisms based on **phase separation, adsorption** and **enthalpic partition** of macromolecules. In the case of charged macromolecules, we encounter also **ion effects,** while **bio-affinity** is to be considered for many biopolymers. Recent developments of polymer HPLC include controlled combinations, **coupling,** of two and more retention mechanisms within the same chromatographic column.

At present, the most often used coupled procedures of polymer HPLC suppress separation of macromolecules according to their molar masses under isocratic and isothermal conditions. The well-known polymer HPLC working under "critical conditions" [8-10] utilizes mutual compensation of entropic (exclusion) and enthalpic contributions to the Gibbs function (ΔG).

In the situation where the resulting $\Delta G = 0$, macromolecules elute from the HPLC column irrespective of their molar mass and the differences in their chemical structure or physical architecture can be assessed [8-12]. The idea of **"critical polymer HPLC"** is very attractive, however, its practical utilization in the area of high molar masses over 100 kg.mol^{-1} is hampered by several recently reviewed experimental problems. [13]

Further coupled procedures of polymer HPLC apply elution of samples in a gradient. One can gradually change either overall eluent composition or temperature. In the former case, polymer species can be separated according to their chemical structure or physical architecture independently of their molar mass. [14-16] In the latter case, highly selective separation of macromolecules according to their molar mass can be achieved [17] and polymer species can also be discriminated according to their nature. [18] Though both eluent and temperature gradient polymer HPLC methods produce interesting results, technical problems, for example those connected with the gradient control and with the quantitative interpretation of results so far complicate their wider application. Therefore, new isocratic methods are still looked for, which could lead to efficient enhancements of polymer separation selectivity according to one of the molecular characteristics of polymers, especially according to their molar masses.

New approach presented in this paper isocratically combines two retention mechanisms. One of them is the conventional entropic partition (size exclusion) of macromolecules, between two liquid phases of the same composition situated in the interstitial volume and in the pore volume. Another retention mechanism is the enthalpic partition of macromolecules between two chemically different phases, namely between the mobile phase (eluent) and the *volume* of stationary (bonded) phase within column packing. The enthalpic partition of polymer species is accompanied by large entropic effects because macromolecules change their conformation as result of energetic interactions. So far, little is known about the latter processes and they will be neglected in this present qualitative discussion.

The enthalpic partition is directly connected with the solubility differences of polymer species in the stationary and in the mobile phases [7]. If macromolecules are (much) better soluble in the mobile phase than in the stationary phase, enthalpic partition mechanism is hardly operative. Macromolecules do not tend entering bonded stationary phase, even though they are small enough to permeate the packing pores. The stationary (bonded) phase, occupies a relatively large fraction of pore volume and therefore the SEC separation selectivity of such systems is reduced. If the sample is similarly soluble in the mobile phase and in the solvated stationary phase, difference is small between retention volumes obtained with the bare and bonded but otherwise identical column packing because the contribution of enthalpic partition is unimportant. If, however, solubility of macromolecules is (much) higher in the solvated stationary phase than in the mobile phase, macromolecules are "pushed" into and retained by the stationary phase. Consequently, their retention volumes rise. Sample may be even fully retained within stationary phase [19] including that situated on the surface of nonporous particles. [20]

As rule, solubility of macromolecules decreases with their increasing molar mass provided the end-group effect is negligible. The actual contribution of enthalpic partition to variation of HPLC retention volumes with polymer molar masses depends on particular system that is on the different effect of polymer size on the solubility reduction in the mobile phase compared to the stationary phase. If solubility of polymer species drops with their M faster in the mobile phase than it does in the solvated bonded phase, the contribution of enthalpic partition to retention volumes rises with sample molar mass. In this way the enthalpic partition acts antagonistically to the size exclusion retention mechanism. On the contrary, if solubility of polymer species drops with M faster in the stationary phase than in the mobile phase, the contribution of enthalpic partition to the overall V_R diminishes with the sample molar mass. Retention volumes, which were increased by the enthalpic partition in the lower molar mass area would be rapidly reduced with M in the higher molar mass range. In this case, the changes in extent of enthalpic partition may "assist" the size exclusion retention mechanism.

The extent of enthalpic partition most likely decreases with the sample molar mass also if macromolecules become too large to fully enter the pores of packing. Under appropriate experimental conditions, the extent of enthalpic partition may rapidly drop with the increasing size of macromolecules. Here again, the enthalpic partition of macromolecules may support the size exclusion retention mechanism. The overall separation selectivity would strongly increase if the reduction of enthalpic partition appears in the narrow range of molar masses. One speaks about **enthalpic partition assisted size exclusion chromatography (EPA SEC).** The introductory study on EPA SEC is presented in this paper.

Experimental

The HPLC apparatus consisted of the pump Model 510 (Waters, Milford, MA, USA) operated at 1 mL min^{-1}, the manual sample injection valve Model 7725 (Rheodyne, Cotati, CA, USA) provided with the sample loop of 50 μL and the evaporative light scattering detector DDL-21 (Eurosep, Cergy-Saint-Pontoise, France). Polymer samples dissolved in the given eluent at the concentration of 1 mg mL^{-1} were injected. Relatively large sample volume was applied into columns of small size because of both limited detector sensitivity and necessity to work with rather low polymer concentrations in order to maintain low viscosity of injected solutions. As known, high viscosity of polymer containing samples causes shifts, broadenings and deformations of solute zones. Column temperature was kept in most experiments at 30 ± 0.01 °C using a custom made column-oven with a duplex wall connected to a water thermostat. The pump was operated at ambient temperature but both eluent and sample solutions were pre-thermostated. Other temperatures applied are given in Figure 7. The data were processed with help of the software Chroma (Chromtech, Graz, Austria).

The retention behavior of several different well and poorly endcapped silica C-18 phases described in [21,22] was evaluated. The results were similar for different columns and only data obtained with Aquasil C-18, 10 nm pores, 5 μm particle sizes (Thermo Hypersil-Keystone, Bellefonte, PA, USA) and Kromasil C-18, 100Å (10 nm pore size), 5 μm particle size, (Eka Chemicals, Bohus, Sweden) are presented and compared in this paper. Column sizes were 250 x 4.6 mm, and 300 x 7.8 mm, respectively. For comparison, selected results are reported with bare silica gel Kromasil 100 Å, 10 μm, column size 300x7.8 mm. Aquasil column is the commercial product. Kromasil sorbents were slurry packed in this Laboratory. The efficiencies of later two columns, measured with toluene injected into tetrahydrofuran eluent applying refractive index detector, were 29,000 theoretical plates.m^{-1} for bare Kromasil and 16,000 for Kromasil C-18. The column efficiencies for eluents containing viscous dimethylformamide are expected to be lower.

Analytical grade solvents were used as eluents, or eluent components viz. tetrahydrofuran from Merck, Darmstadt, Germany, and dimethylformamide from Scharlau, Barcelona, Spain. They were vacuum distilled before use. Tetrahydrofuran was treated with KOH and Na before distillation and the distilled solvent was stabilized with 0.02% of butylated p-cresol. Mixed eluents were prepared by weighing and the control of eluent composition was better than ± 0.1 wt. %.

Two sets of homopolymers differing in their polarities were applied. They exhibited narrow to medium molar mass distributions. In all cases, the peak retention volumes could be well identified. Polystyrenes (PS) were from Pressure Chemicals Co., Pittsburgh, PA, USA (molar masses ranged from 0.666 to 1,200 kg.mol^{-1}), and poly(methyl methacrylate)s (PMMA) of low stereoregularity were gifts from Dr. W. Wunderlich, Röhm, Darmstadt, Germany [23] and from Dr. J. Herz of Institut Sadron, Strasbourg, France (M ranged from 1.3 to 613 kg.mol^{-1}).

A set of block copolymers PS-b-PMMA was also measured. They were prepared by anionic polymerization. The molar masses of blocks were nearly equal and assumed values about 10, 30, 50 and 90 kg.mol^{-1} (Polymer Standards Services, Mainz, Germany). After each set of experiments the retained macromolecules were removed from columns by an overnight action of the efficient displacer, THF. Columns were re-equilibrated by the fresh eluent before the next series of measurements.

Results and Discussion

To avoid problems with mixed enthalpic retention, the adsorption of macromolecules on the column packing surface must be suppressed. Recently, it was found [21,22] that some medium-, and high-polarity polymers were adsorbed from the low- or medium-polarity eluents on the free silanols, which are abundant on the surface of most commercial silica C-18 bonded phases. For very polar polymers, the extensive adsorption was observed even with the carefully end-capped silica C-18 column packings.[22] However, the adsorption of low polarity macromolecules such as polystyrenes on the end-capped silica C-18 phases is improbable. Moreover, polar dimethyl formamide (DMF) and medium polar tetrahydrofuran (THF), which suppress adsorption of medium polarity macromolecules on silica gel, were applied in our study as eluents. DMF even fully prevented adsorption of high polarity polymers such as poly(2-vinyl pyridine)s and polyethylene oxides on free silanols of bare and bonded silica gels at its concentration of about 30% in a mixed eluent DMF/THF. [22] The low polarity macromolecules of PS are adsorbed on bare silica gel only from non-polar solvents like cyclohexane or carbon tetrachloride. [24,25] Adsorption of PS on bare and bonded silica gels was not observed from medium and low polarity, eluents such as THF and toluene, respectively. We can conclude that the adsorption of PS on the silica gel C-18 column packings from DMF and THF is hardly possible.

The exponents in Kuhn-Mark-Houwink-Sakurada viscosity law, a, for PS in THF at 25-30 °C range from 0.64 to 0.768. [26] This large scatter is explained mainly by the high hygroscopicity of THF, which rapidly absorbs substantial amount of water and it is anticipated that the proper a value for PS in dry THF lies in the area of 0.73. [26] It means that THF is a good solvent for PS, likely much better than the solvated C-18 groups of the bonded phase. Therefore THF should suppress enthalpic partition of polystyrene in favor of the C-18 phase. The dependences of log V_h vs. V_R for PS standards in THF eluent monitored for bare silica gel and for the C-18 phase prepared from the same starting silica gel are compared in Figure 1.

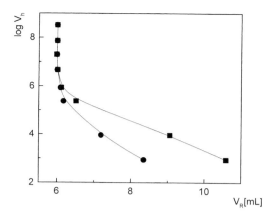

Figure 1. Dependences of log V_h vs. V_R for bare silica gel (■) and for silica gel C-18 phase (●) prepared from the same carrier, Kromasil 100. Polystyrenes were injected in a good solvent, tetrahydrofuran. The curve for silica gel C-18 was normalized considering excluded retention volume (V_0) for bare silica gel. For discussion see the text.

V_h is the hydrodynamic volume of macromolecules [27] expressed: $V_h = M [\eta]$, where M is the most abundant molar mass in the sample and $[\eta]$ is the corresponding limiting viscosity number in eluent. The evaluation of courses of these dependences for the same column but with various polymers and various eluents allows assessment of enthalpic interactivity of HPLC columns [19,21,22] and at least semi-quantitative determination of averages and distributions of packing pore sizes. [28,29] The difference in the shapes of the dependences in Figure 1 is obvious. It reflects unequal pore volume accessible for PS molecules in bare and bonded silica gel. Both the excluded V_h values and the retention volumes of smaller polymer species decreased due to presence of the C-18 groups. It scan be concluded that molecules of PS hardly enter the C-18 bonded phase in THF and both the effective pore size and pore volume become smaller. In a reasonable approximation, enthalpic interactions can be neglected and entropic partition of macromolecules can be considered the only retention mechanism present in this system ("**ideal SEC**"). The course of dependence of log V_h vs. V_R for bare Kromasil also indicates relatively low pore volume of this material (see also [21]), which is therefore generally less suitable for SEC separations. It is supposed that the effective volume of the C-18 phase can be estimated from dependences shown in Figure 1. [19]

DMF is a rather poor solvent for PS. The exponent a at 30 °C is 0.612. [30] DMF strongly promoted enthalpic partition of PS species in favor of C-18 phases [7,19] and molar masses higher than about 1 kg.mol^{-1} were even fully retained within C-18 phases bonded to porous silica. [20]

Basing on the above observations, THF has been chosen as the partition suppressing eluent component and DMF as the partition promoting eluent component for polystyrenes in connection with the silica C-18 phases. It was anticipated that by mixing these two solvents it would be possible to finely control extent of PS partition. The dependences of log M vs. V_R (further only "the Plots") for Aquasil C-18 are depicted in Figure 2.

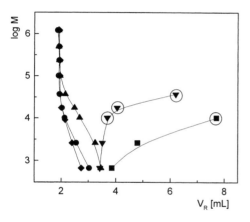

Figure 2. The Plots of log M vs. V_R for polystyrenes eluted from Aquasil C-18 column packing in DMF (■), THF (◆), and in mixed eluents DMF/THF containing 90 (▼), 80 (▲) and 50 (●) wt. % of DMF. The data points for systems exhibiting reduced apparent sample recovery are marked with additional circles. For explanation see the text.

Aquasil contains "embedded" polar groups within the C-18 phase. Polar groups are introduced into some C-18 bonded phases in order to control their polar interactivities and to prevent their collapse in the HPLC eluents containing high concentration of water. The producer did not disclose nature, amount and position of embedded polar groups. The overall retention of polystyrenes in DMF on Aquasil was dominated by enthalpic partition and the retention volumes increased with M. Molar masses over 10 kg mol^{-1} were fully retained within column. Addition of THF to eluent suppressed enthalpic partition and the character of the Plots in Figure 2 changed from the enthalpic partition dominated to the exclusion dominated one for the eluents containing between 10 and 20 wt. % of THF. Addition of 50 wt. % of THF to DMF brought the Plot almost to the coincidence with that for pure THF. Similar courses of the Plots were observed for Kromasil C-18 with different DMF/THF mobile phases compositions (Figure 3), except for the fact that higher molar masses of PS have been eluted from Aquasil in the pure DMF eluent compared to Kromasil C-18. [19] This indicates that presence of the embedded polar groups slightly suppressed enthalpic partition of PS species in favor of the C-18 phase.

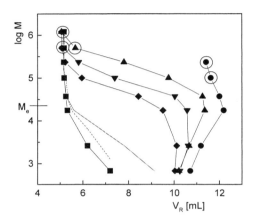

Figure 3. The Plots of log M vs. V_R for PS eluted from Kromasil C-18 with THF ($\blacksquare$) and with mixed eluents DMF/THF containing 83 ($\bullet$); 82 ($\blacktriangle$); 81 ($\blacktriangledown$), and 80 ($\blacklozenge$) wt. % of DMF. For comparison also the Plots are shown for PMMA in DMF/THF with 80 wt. % of DMF (- - -) (compare also Figure 5), and for PS in THF on bare Kromasil (—) (normalized to $V_0 = 5.17$ mL). M_e is the molar mass of PS excluded from the silica C-18 packing in THF that is under SEC conditions. The reduced apparent sample recovery is depicted with additional circles. The sample recovery for eluent containing 80 wt. % of DMF approached 100 %.

Small changes in the eluent composition around 80 wt. % of DMF strongly affected the courses of the Plots in Figure 3. Evidently, the extent of enthalpic partition of polystyrene between mobile phase and solvated bonded C-18 intensively varied in the vicinity of this eluent composition. The courses of the Plots around 80 wt.% of DMF reflected highly selective SEC-like separation of polystyrenes, which extended only over about 1.5 order of magnitude in M. This is less than the expected highest theoretical selectivity of separation based solely on size exclusion. The SEC column packings with pores of uniform size should produce the Plots extending over two orders of magnitude in M. [1] The excluded molar masses in eluents containing 80 to 82 wt. % of DMF fairly exceeded M_e for the silica C-18 in THF and even that for the bare silica gel. Similarly, retention volumes of polymers with the lowest M were higher than that found for silica C-18 phase, as well as for identical bare silica gel packing. This behavior can be attributed to the effect of the enthalpic partition of polystyrenes in favor of the C-18 phase, which assisted the SEC retention mechanism.

Notice the unexpected back-turn course of the Plots for eluents containing 81 and, especially 82 wt. % of DMF in Figure 3. These Plots are likely composed of two different parts. The enthalpic partition slightly prevailed up to about the SEC excluded molar mass of PS, while above this limit the exclusion process clearly dominated. Similar back-turn courses of the Plots have been observed with several other systems, in which enthalpic partition was combined with exclusion. [7,19] It is evident, that the experimental conditions leading to the back-turn shaped Plots must be avoided in most chromatographic systems applied for analytical purposes.

The high selectivity separation of macromolecules assisted by enthalpic partition (EPA SEC) is documented with a mixture of three polystyrenes in Figure 4.

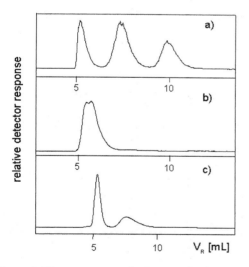

Figure 4. Chromatograms of mixtures of polystyrenes with M 100, 37 and 9 kg.mol⁻¹ obtained a) with Kromasil C-18 column 300x7.8 mm and mixed eluent DMF/THF containing 80 wt. % of DMF; b) with the same column as a) and pure THF eluent; c) with bare Kromasil column 300x7.8 mm and pure THF. Interstitial volume for bare silica gel column was slightly higher than that for C-18 column.

The base line separation for polystyrenes with similar molar masses has been easily achieved with one single column in spite of its rather low efficiency (see the Experimental section) and a relatively high viscosity of eluent. For comparison also chromatograms are shown in Figure

4, which were obtained without presence of enthalpic partition. The same mixture of polystyrenes was eluted from bare silica gel and C-18 bonded silica gel using pure THF eluent. Selectivity of separation in the latter cases was much lower and this demonstrates the result of enthalpic partition "assistance".

The Plots of log M vs. V_R obtained for poly(methyl methacrylate)s with the same column and with eluents of similar composition as in Figure 3 are shown Figure 5.

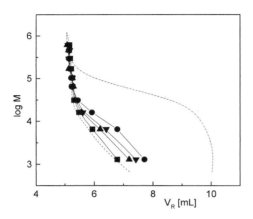

Figure 5. The Plots of log M vs. V_R for PMMA eluted from Kromasil C-18 in THF (■), and in DMF (●), as well as in mixed eluents DMF/THF containing 80 (▲) and 81 (▼) wt. % of DMF. For comparison also the Plots for PS in THF (- - -) and in the mixed eluent containing 80 wt. % of DMF (-.-.-) are depicted.

DMF is only a little better solvent for PMMA (a = 0.625 [31]) than for PS. Still, it seems that the enthalpic partitioning of more polar PMMA in favor of solvated C-18 groups is much less pronounced than in the case of less polar PS in DMF and DMF containing eluents. The enthalpic partition of PMMA appears only in the "SEC non-excluded" molar mass area. This gives a chance for separation of less polar polymers such as PS from the more polar species such as PMMA - even if their molar masses are similar: PMMA samples with M above 30 kg.mol^{-1} will leave the column in the excluded retention volume V_0 while less polar macromolecules of PS will elute much later, under the EPA SEC conditions. An example of such approach is demonstrated in Figure 6.

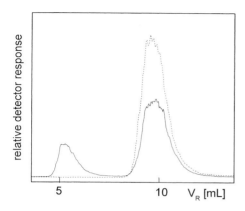

Figure 6. Chromatograms of the mixture of PS (17.5 kg.mol^{-1}) with PMMA (16 kg.mol^{-1}) (—), and of the same single PS (- - -). Column Kromasil C-18, 300x7.8 mm, mixed eluent DMF/THF containing 80 wt. % of DMF.

Polystyrene was separated from poly(methyl methacrylate) possessing almost identical molar mass. The chromatogram of PS remained practically unaffected by presence of PMMA. The PMMA species leaving an EPA SEC column can be subsequently characterized by an appropriate on-line SEC column, which does not exhibit enthalpic partition that is in a two-dimensional polymer HPLC arrangement.

In Figure 7, the effect of temperature on the courses of the Plots is demonstrated for Kromasil C-18 with the eluent containing 80 wt. % of DMF. The flow rate was not corrected for thermal contraction/expansion of eluent. The corrections would not affect tendencies in the courses of the Plots. The Plots changed their shapes from enthalpic partition dominated to size exclusion (entropy) dominated retention with increasing temperature. The system was very sensitive to temperature change in the range from 10 to 25 °C, and especially between 11.5 and 13 °C. The back-turn shaped Plots were observed between 12.7 and 15 °C. The solubility of PS likely improved with temperature faster in eluent than in the solvated C-18 phase and, therefore, the overall extent of enthalpic partition of PS dropped with rising temperature. It is evident that the molar mass range, in which the EPA SEC selective separation takes place, can be easily adjusted by changing the column temperature. A tandem of columns packed with the same material but working at different temperatures can cover a rather broad molar mass area. This is experimentally well feasible, however, more attractive is the opportunity to ad hoc create a tailored EPA SEC system, which would highly selectively discriminate samples just in the desired molar mass area. It is anticipated that the temperature adjustment will allow to extend the working area of the EPA SEC method to the molar masses well exceeding thousands of kg.mol-1 with relatively narrow pore column packings of less than 100 nm diameter. The work with mechanically instable ultra-wide pore SEC column packings could be avoided in this way. However, one must consider the back-turn courses of some Plots, as well as the possible reduction in the sample recovery (see below). It is not yet clear if the EPA SEC principle will be applicable to oligomers. The method apparently works well only in the domain of molar masses excluded from the given packing under "ideal" SEC conditions. As known, the temperature dependences of retention volumes allow discrimination of enthalpic and enthalpic contributions to polymer retention. The quantitative evaluation, which needs large series of very exact data is under preparation.

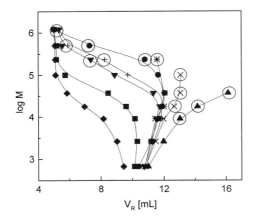

Figure 7. The Plots log M vs. V_R for PS eluted from Kromasil C-18 with mixed eluent DMF/THF containing 80 wt. % of DMF at different temperatures (▲) 10 °C; (×) 11.5 °C; (✳) 12.3 °C; (●) 12.7 °C; (+) 13 °C; (▼) 15 °C; (■) 25 °C, and (◆) 50 °C. The data points for systems exhibiting reduced apparent sample recovery are marked with additional circles.

An important issue for all polymer HPLC techniques, which couple entropic and enthalpic retention mechanism represents sample recovery. This is the case for polymer HPLC under critical [13] and limiting [32] conditions – and also for EPA SEC. In all above methods, macromolecules may exhibit a tendency to stay retained within column irrespectively of the amount of mobile phase pumped through the system ("full retention"). The full retention problem augments with decreasing diameter of the packing pores and with increasing both sample molar mass and enthalpic interaction between macromolecules. Selective full retention of high molar mass fractions affects results of analyses also indirectly because the sample fractions retained within column alter both the effective pore size and the overall interactivity of packing.

The low sensitivity of evaporative light scattering detector (ELS) used in this study, as well as intrinsic limitations of ELS detection in general - the non linear response toward sample concentration [33,34], and the dependence of response on the eluent nature [34] - prevented quantitative evaluation of sample recovery. The "apparent" sample recovery was considered "decreased" when the peak area monitored by the ELS detector for the EPA SEC system dropped more than 25% when compared to the "ideal" SEC conditions. In this case, data points in Figures are marked with an additional circle (Figures 2, 3 and 7).

The exact molecular backgrounds of processes taking place in the EPA SEC systems are not well understood, yet. One can speculate that the **difference** of solubilities of polymer molecules in the stationary and in the mobile phase rapidly changes in the area of M subject to selective separation. As mentioned, solubility of macromolecules may decrease much faster in the bonded phase than in the mobile phase with rising molar mass. As result, the extent of enthalpic partition would decrease with increasing polymer molar mass. This hypothesis, however, does account neither for the increased excluded molar mass of samples in EPA SEC compared to an identical "ideal" SEC column nor for the back-turn shape of some EPA SEC Plots. Further, there is no apparent reason for abrupt change in the solubility difference to appear just at a relatively high sample molar mass nearly corresponding to that excluded from the packing pores in the "ideal" SEC separation mode.

The alternative explanation of the EPA SEC phenomenon supposes that the enthalpic interactions push also large (the SEC excluded) polymer species into the packing pores. [7] This process intensifies with increasing attractive segmental interaction energy between sample molecules and column packing, ε, which is sensitive to temperature, eluent composition and possibly also to pressure. At very high ε values, and for eluents, which are thermodynamically poor for the polymer under study (low a values in the viscosity law) enthalpic partition in favor of stationary phase may be so large that macromolecules of practically any molar mass stay retained within column and cannot be eluted by any volume of eluent (full retention). It is supposed that both the attractive interactions between packing and macromolecules and the repulsive interactions between eluent and macromolecules force parts of the de-coiled macromolecules to enter even the narrowest pores to create "stems" of the polymer species assuming a "flower like" conformation. [35-37] Under such conditions, macromolecules may be attached to the column packing simultaneously with many adjacent segments. This attachment may be very strong and difficult to be cancelled by molecules of a displacing liquid [32], also due to slow processes of diffusion.

As the attractive interactions packing – polymer (ε) decrease and the attractive interactions polymer – eluent (a) increase, the retention volumes decrease. At the same time, ever larger macromolecules start eluting (see for example Figure 2). Eventually, we arrive at the above mentioned "critical conditions" where the entropic partition (exclusion) and the enthalpic partition mutually compensate and polymer retention volumes become independent of the sample molar mass. [8-10] The plots of $log\ M$ vs. V_R are vertical. It seems that in the vicinity of critical conditions, the flower like interactions are still feasible. This means that at least a fraction of macromolecules, which are excluded at $\varepsilon \sim 0$ and at high a (the SEC system) may enter some packing pores at $\varepsilon > 0$ and at low a. The dept of such enthalpy forced pore permeation probably rapidly decreases with the sample molar mass. Therefore, the retention volume augmentation due to enthalpic partition sharply drops with the increasing sizes of macromolecules. This process can be responsible for the EPA SEC behavior of macromolecules.

Clearly, the overall retention of macromolecules is dominated by enthalpic partition (V_R's increase with M) when ε exceeds its critical value. Above certain molar mass, however, the effect of enthalpic partition starts decreasing with M, anyway, and so do the retention volumes. As result, the back-turn courses of the Plot appear. In this case, the area of "enthalpic assistance" to exclusion retention mechanism is shifted to the higher sample molar masses.

It is anticipated that the selectivity of the SEC separation of polymers can be augmented also by the adsorption processes. [7,38] The resulting procedure can be termed **adsorption assisted size exclusion chromatography.**

An interesting application of EPA SEC would be the characterization of block copolymers. A chromatographic system can be identified, in which one of the blocks (for example PMMA) remains non-retained (excluded) while another block (for example PS) will be eluted under EPA SEC conditions. In this way, the PS blocks would be characterized irrespective of presence of PMMA blocks. This idea was tested with a series of PS-b-PMMA samples. Contrary to expectations, elution behavior of block copolymers was governed by the PMMA blocks. Samples of block copolymers eluted in the excluded volume of column in the form of rather narrow peaks or at the V_R of PMMA blocks (results not shown). It seems that the presence of PMMA blocks in diblock copolymers PMMA-block-PS prevented enthalpic partition of PS chains. The interpretation of these results will, however, need further experiments. This last observation may also shed light on the behavior of diblock copolymers in the "critical polymer HPLC", where some deviations from expected behavior were found by Chang et al. [39]

Conclusions

A new member was added to the family of methods of polymer HPLC namely enthalpic partition assisted size exclusion chromatography (EPA SEC). EPA SEC combines enthalpic and entropic partition of macromolecules within the HPLC column to allow a selective separation of polymer species according to their molar mass or chemical nature/composition. Enthalpic partition of macromolecules within appropriate stationary phases is promoted by the low solubility of polymer species in eluent. The extent of enthalpic partition is controlled by the amount of a thermodynamically good solvent added to the mobile phase. The additive suppresses extent of enthalpic partition. Several systems polymer-eluent-column packing, which exhibited excessive enthalpic partition were identified in the studies initially devoted to the tests of enthalpic retentivities of HPLC column packings [19,21,22] and to the evaluation of enthalpic retention mechanisms of macromolecules. [7] These systems included polystyrenes (PS) and poly(n-butyl methacrylate)s (PnBMA) model polymers, dimethylformamide (DMF) and diethylmalonate (DEM) as enthalpic partition promoting solvents, as well as tetrahydrofuran (THF) and toluene as enthalpic partition suppressing liquids – all in combination with a series of different silica C-18 bonded phases. Only few typical results obtained with polystyrenes, DMF/THF mixed eluents and with two different silica C-18 column packings are presented in this paper, however, the behavior of PnBMA polymers and DEM/THF and DEM/toluene mixed eluents was very similar to that reported here. Therefore the generalization of basic conclusions is possible. The tentative explanation of high selectivity separation of homopolymers by EPA SEC assumes

- a presence of enthalpic partition of macromolecules in favor of the (bonded) stationary phase. Enthalpic partition results from better solubility of macromolecules in solvated stationary phase than in the mobile phase

- an increase of polymer retention volumes caused by enthalpic partition compared to that due to the sole entropic partition (exclusion)

- a fast decrease of enthalpic partition of macromolecules with their increasing molar masses under specific experimental conditions. The sharp drop of enthalpic partition appears above certain value of sample molar mass, M_{ep}, which as rule lies in the vicinity of polymer molar mass fully excluded from the packing pores in the "ideal" size exclusion chromatographic mode (M_e). For a given polymer and column packing the actual value of M_{ep} depends on eluent composition and/or on temperature of experiment.

The mechanism behind EPA SEC may include the molar mass dependent

- pushing the parts of large (SEC excluded) polymer chains by enthalpic interactions into the solvated bonded phase even if the latter is situated in the narrow pores of column packing

- differences between solubilities of macromolecules in the stationary phase and in the mobile phases.

EPA SEC can be used for selective size separations of homopolymers within narrow molar mass areas, as well for discrimination of polymer blend components. Further applications of EPA SEC may include separation of macromolecules according to their composition and architecture. The efficacy and feasibility of EPA SEC must be, however, checked with a series of real systems and with the homogeneous bonded phases of different volumes and polarities. The effects of experimental conditions, for example packing pore size, injected polymer concentration and volume, flow rate and pressure in the system, as well as column temperature must be carefully evaluated together with the role of sample recovery in order to decide about practical applicability of method.

Acknowledgements

This work was supported by the Slovak Grant Agency Vega, project 2-301 123. The author thanks Drs. W. Wunderlich and J. Herz for samples of poly(methyl methacrylate)s, Thermo Hypersil and Eka Chemicals (Akzo Nobel) for the gift of columns, as well as Mrs. J. Tarbajovska for her excellent technical assistance.

[1]   W. W. Yau, S. R.Ginnard, J. J.Kirkland, *J. Chromatogr.* **1978**, *149*, 465.
[2]   E. F. Casassa, *J. Polym. Sci. B* **1967**, *5*, 773.
[3]   E. F. Casassa, *J. Phys. Chem.* **1971**, *75*, 3929.
[4]   A. A. Gorbunov, A. M. Skvortsov, *Adv. Colloid Interface Sci.* **1995**, *62*, 51.
[5]   I. Teraoka, *Progr. Polym. Sci.* **1996**, *21*, 89.
[6]   P. Cifra, T. Bleha, *Polymer* **2000**, *41*, 1003.
[7]   D. Berek, *Chromatographia*, Supplement **2003**, *57*, S-45.
[8]   B. G. Belenkii, M. D. Valchikhina, I. A. Vakhtina, E. S. Gankina, O. G. Tarakanov, *J. Chromatogr.* **1976**, *129*, 115.
[9]   M. B. Tennikov, P. P. Nefedov, M. A. Lazareva, S. Y. Frenkel, *Vysokomol. Soedin. (Moscow) Ser. A* **1977**, *19*, 657.
[10]   A. M. Skvortsov, A. A. Gorbunov, *Vysokomol. Soedin. (Moscow) Ser. A* **1980**, *22*, 2641.
[11]   M. Janco, T. Hirano, T. Kitayama, K. Hatada, D.Berek, *Macromolecules* **2000**, *33*, 1710.

[12] S. G. Entelis, V. V. Evreinov, A. V. Gorshkov, *Adv. Polym. Sci.* **1987**, *76*, 129.
[13] D. Berek, *Macromol. Symp.* **1996**, *110*, 33.
[14] S. Teramachi, A. Hasegawa, Y. Shima, M. Akatsuka, M. Nakajima, *Macromolecules* **1979**, *12*, 992.
[15] T. H. Mourey, *J. Chromatogr.* **1986**, *357*, 101.
[16] H. Sato, K. Ogino, T. Darwint, I. Kiyokawa, *Macromol. Symp.* **1996**, *110*, 177.
[17] T. Chang, H. C. Lee, W. Park, C. Ko, *Macromol. Chem. Phys.* **1999**, *200*, 2188.
[18] H. C. Lee, T. Chang, *Macromolecules* **1996**, *29*, 7294.
[19] D. Berek, *J. Chromatogr. A.* **2003**, *1020*, 219.
[20] S. H. Nguyen, D. Berek, *Int. J. Polym. Anal. Charact.* **2002**, *7*, 52.
[21] D. Berek, *J. Chromatogr. Part A.* **2002**, *950*, 75.
[22] D. Berek, J. Tarbajovska, *J. Chromatogr. Part A.* **2002**, *976*, 27.
[23] D. Berek, M. Janco, T. Kitayama, K. Hatada, *Polym. Bull.* **1994**, *32*, 629.
[24] C. Van der Linden, R. Van Leemput, *J. Colloid Interface Sci.* **1978**, *67*, 63.
[25] L. Litvinova, N. Bolnikevich, E. Tyihak, *J. Planar Chromatogr.* **2000**, *13*, 149.
[26] T. Spychaj, D. Lath, D. Berek, *Polymer* **1979**, *20*, 437.
[27] Z. Grubisic, P. Rempp, H. Benoit, *J.Polym.Sci.* **1967**, *B5*, 573.
[28] S. B. Schram, D. H. Freeman, *J. Liq. Chromatogr.* **1980**, *3*, 403.
[29] R. N. Nikolov, W. Werner, I. Halasz; *J. Chromatogr. Sci.* **1980**, *18*, 207.
[30] M. Zinbo, J. L. Parsons, *J. Chromatogr.* **1971**, *55*, 55.
[31] I. Kossler, M. Netopilik, G. Schulz, R. Gnauck, *Polym. Bull.* **1982**, *7*, 597.
[32] M. Snauko, D. Berek, *Chromatographia* Supplement **2003**, *57*, S-55.
[33] A. M. Striegel, *J. Chromatogr. Part A* **2002**, *971*, 151.
[34] B. Trathnigg, M. Kollroser, D. Berek, S. H. Nguyen, D. Hunkeler, in: Provder, T., editor „Chromatography of Polymers", *ACS Symp. Ser.* **1999**, *731*, 95.
[35] A. M. Skvortsov, L. T. Klushin, A. A. Gorbunov, *Macromolecules* **1997**, *30*, 1818.
[36] A. M. Skvortsov, A. A. Gorbunov, F. A. M. Leermakers, G. J. Fleer, *Macromolecules* **1999**, *32*, 2004.
[37] E. Zhulina, A. V. Dobrynin, M. Rubinstein, *Eur. Phys. J., Part E* **2001**, *5*, 41.
[38] D. Berek, M. Janco, G. R. Meira, *J. Polym. Sci. A: Polym. Chem.* **1998**, *36*, 1363.
[39] W. Lee, D. Cho, T. Chang, K. J. Hanley, T. P. Lodge, *Macromolecules* **2001**, *34*, 2353.

Macromol. Symp. **2004**, *216*, 165-178

Study on the Kinetics for Enzymatic Degradation of a Natural Polysaccharide, *Konjac Glucomannan*

Guangji Li,[*1] *Li Qi,*[1] *Aiping Li,*[1] *Rui Ding,*[1] *Minhua Zong*[2]

[1] Department of Polymer Science and Engineering, South China University of Technology, Guangzhou 510640, P. R. China
E-mail: guangjili@hotmail.com
[2] Biotechnology Department, South China University of Technology, Guangzhou 510640, P. R. China

Summary: The enzymatic degradation of konjac glucomannan (KGM) was conducted using β-mannanase from an alkalophilic *Bacillus sp.* in the aqueous medium (pH 9.0) at 30°C. The intrinsic viscosity ($[\eta]$), molecular weight (M_w) and molecular weight distribution (MWD) of the degraded KGM were measured. The mathematical relation between $[\eta]$ and M_w, $[\eta] = 5.06 \times 10^{-4} M_w^{0.754}$, was established. The kinetic analysis reveals a dependence of the rate constant (k) on the period of reaction and the initial substrate concentration (c_0) over the range of substrate concentration (1.0~2.0%) used in this work. The results indicate that the enzymatic degradation of KGM is a complex reaction combining two reaction processes with different orders. In the initial phase of degradation k is inversely proportional to c_0, which is characteristic of a zeroth-order reaction; while in the following phase k is independent of c_0, implying the degradation follows a first-order reaction. The reactivity difference in breakable linkages of KGM, the action mechanism of an enzyme on KGM macromolecules, and the theory concerning the formation of an enzyme-substrate complex and 'substrate saturation' can be used to explain such a kinetic behavior. In addition, the enzymatic degradation of KGM was also carried out using the other enzymes like β-mannanase from a *Norcardioform actinomycetes*, β-glucanase Finizym and a compound enzyme Hemicell as a biocatalyst. By comparing and analyzing the degradation processes of KGM catalyzed by four different enzymes, it can be observed that there is a two-stage reaction with two distinct kinetic regimes over a certain range of degradation time for each of the degradation processes. These results are useful to realize controllable degradation of polysaccharides via an environmental benign process

Keywords: degradation; enzymes; kinetics; konjac glucomannan; polysaccharides

DOI: 10.1002/masy.200451216

Introduction

In recent years green chemicals and materials, as well as safe and reliable medicines, have drawn much attention due to the present bad environmental situation caused by overburdensome pollutants and due to the increasing anxiousness about public health caused by the production and use of a large number of synthetic chemicals and medicines. On the other hand, developing and utilizing renewable resources have become a very important policy in many developed and developing countries to decrease the demand for and the dependence on limited or depleting resources like petroleum and to realize the goal of sustainable development. As renewable resources, naturally occurring polysaccharides have been revalued because they are environmentally friendly and they possess great potentials for industrial, agricultural, and medical applications as substitutes for some petrochemical products.[1] Konjac glucomannan (KGM) is a kind of natural polysaccharide and renewable resource from the *Amorphophallus konjac* plant. KGM and its derivatives have aroused a great interest of some reasearchers because of their biological activity and many unique physiological and pharmacological functions. It is composed of D-mannose and D-glucose chains linked by β-1, 4-pyranoside bonds; and the molar ratio of mannose and glucose is 1.6 to 1 or 1.69 to 1, depending on the origin of KGM.[2, 3] There are certain short side branches at the C-3 position of the mannoses and acetyl groups randomly present at the C-6 position of a sugar unit.[4, 5] The acetyl groups frequently range from one per six sugar units to one per twenty sugar units.[5] KGM has been widely used in the food, chemical, textile, oil and cosmetics industry because of its unique rheological properties.[6, 7] It is worth noting that KGM or its derivatives may be attractive for biomedical applications since they possess some interesting biological or immunological activities.[8-10]

However, the molecular weight of original KGM is in the range of 10^6 and its solution exhibits a very high viscosity. When it is degraded to a certain extent, KGM can be modified by means of environmentally benign biocatalysis or chemo-biocatalysis, thus leading to valuable applications. For this reason we focused our study on regularities happening in the enzyme-catalyzed degradation of KGM.[11] Although the enzymatic degradation or hydrolysis of KGM have already been studied,[3, 12-15] these researches mainly involved the nearly complete degradation or hydrolysis of KGM, which aimed to prepare monosaccharides, disaccharides or oligosaccharides, to analyze the structure of KGM and to study the mechanism of this reaction. To our knowledge, the studies on 1 controllable or imited degradation of KGM conducted by means of an environmentally benign biocatalysis reaction performed under mild conditions, are seldom reported.

In the present work, we investigated the degradation kinetics of KGM catalyzed by β - mannanase from an alkalophilic *Bacillus sp.* (Mannanase I) and compared the characteristics of the degradation kinetics of KGM catalyzed by four different enzymes, thus laying a theoretical and experimental foundation for a controllable degradation of KGM and other polysaccharides via an enviroenmentally benign process. Some results are reported here.

Experimental

Materials and chemicals. Original Konjac glucomannan (KGM), purified powder (3A-PF-120), was a gift from Multi-Ring Health Products, Ltd., China. The β -mannanase (EC.3.2.1.78) from an alkalophilic *Bacillus sp.* (Mannanase I) and another β -mannanase from a *Norcardioform actinomycetes* (Mannanase II) were provided by the Institute of Microbiology, Chinese Academy of Science, β -glucanase Finizym was given by Novo Nordisk Corp., Denmark and a compound enzyme Hemicell was furnished by Chemgen Corp., USA. These enzymes were used as a biocatalyst in the degradation of KGM, respectively. The other chemicals were analytical grade reagent and used as purchased.

Enzymatic degradation. The enzymatic degradation of KGM as a substrate was run in a glass batch reactor containing a certain volume of reaction medium with the optimum pH for the used catalyst, stirred in a thermostat water bath at 30°C. The reaction system was appropriately varied according to the characteristics and activity of the enzyme used as a biocatalyst in the reaction, as shown in Table 1.

At a prescribed time, ethanol was added to inactivate the enzyme, leading to the end of a reaction. The degraded KGM product was repeatedly washed and reprecipitated with water and ethanol, then followed by lyophilization.

Table 1. Reaction system for the enzymatic degradation of KGM catalyzed by four different enzymes, respectively.

Biocatalyst	Reaction Medium		Enzyme Contcentration		Substrate Contcentration (%)
	Buffer Solution	pH			
Mannanase I	glycine-NaOH	9.0	8	mg/dL	1.0 ~ 2.0
Mannanase II	Na_2HPO_4-citric acid	8.0	2	mg/dL	1.0
Finizym	Na_2HPO_4-citric acid	5.0	0.025	mL/dL	1.0
Hemicell	Na_2HPO_4-citric acid	7.0	0.0005	mL/dL	1.0

Viscosity measurement. The degraded KGM sample was dissolved in distilled water at 0.05 g/dL. The intrinsic viscosity of the solution was measured using Ubbelohde-type viscometer at $30 \pm 0.5°C$.

GPC Measurement. The molecular weight (M_w) and molecular weight distribution (MWD, or M_w/M_n) of the degraded KGM samples were measured at 40°C by gel permeation chromatography (GPC) using a commercial GPC system equipped with a column (Ultrahydrogel 500, Waters Corp., Milford, MA, USA), a pump (Waters, Model 510) and a differential refractometer (Waters, Model 410). The system was calibrated with a monodisperse polyethylene glycol (PEG) standard. The mobile phase was NaCl aqueous solution. Flow rate was 0.6 mL/min.

Results and Discussion

Change in Intrinsic Viscosity of Enzymatically Degraded KGM Products. Figure 1 illustrates the changes in the intrinsic viscosity $[\eta]$ of the enzymatically degraded KGM samples by Mannanase I with the reaction degradation time t for different substrate concentrations. The $[\eta]$ indirectly reflects the molecular weight of the measured polymer sample. It is obvious that $[\eta]$ decreases rapidly in the initial period of the reaction and then slows down as degradation takes place. The viscosity reduction is greater for the reaction system with a lower substrate concentration.

When the degradation of KGM was catalyzed at the same substrate concentration (1.00 %) by differenrt enzymes, the intrinsic viscosity $[\eta]$ of the enzymatically degraded products was found to change with the degradation time t following the same trend as shown in Figure 2. Withou a doubt, a decrease in the intrinsic viscosity $[\eta]$ is caused by the degradation of KGM. A comprehensive analysis of the structural features of KGM, the characteristics of a biocatalyst and its catalytic mechanism in this reaction is helpful to explain the above-described behavior.

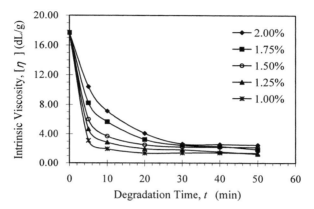

Figure 1. The curves of intrinsic viscosity $[\eta]$ *versus* degradation time t for the enzymatically degraded KGM samples by Mannanase I at different substrate concentrations.

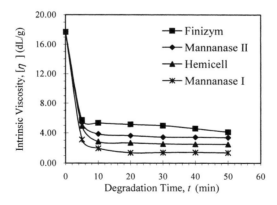

Figure 2. The curves of intrinsic viscosity $[\eta]$ *versus* degradation time t for the enzymatically degraded KGM samples by deferent enzymes at the same substrate concentration, 1.00 %.

Since the macromolecular backbone of KGM is built up of D-mannopyranosyl residue (M) and D-glucopyranosyl residue (G) that are randomly bonded by β-1, 4-glycosidic linkage (→),[12-14, 16] there may exist four different β-1, 4-glycosidic linkages bonding different glycosyl units together: β-mannosidic linkage (M→M), β-glucosidic linkage (G→G), β-mannosyl-glucosidic linkage (M→G) and β-glucosyl-mannosidic linkage (G→M). The studies on the enzymatic and acidic hydrolysis of KGM[12-14, 16] have indicated that theses bonds possess different "strength", which increases following the order M→G, M→M, G→M and G→G. Thus, in the initial period of the enzymatic degradation of KGM the relatively weak bonds, M→G and M→M, can be rapidly broken while the relatively strong links like G→M and G→G remain. As the reaction goes on, the chain length of substrate molecules decreases, leading to a reduction in the susceptible linkages and the active sites of substrate for an enzyme to bind so that the $[\eta]$ or molecular weight of degraded products does not change dramatically.

Establishment of the Relationship between the Intrinsic Viscosity and the Molecular Weight of KGM Samples. The GPC measurements for an original KGM and a series of the degraded KGM samples show that the M_w and MWD of an original KGM are about 9.8×10^5 and 1.46, respectively; and the degraded KGM samples exhibits a wider range of MWD from 1.68 to 1.95. A good linear relationship is observed between $\mathrm{Log} M_w$ and $\mathrm{Log}[\eta]$ for the degraded KGM solutions as shown in Figure 3. Thus, the mathematical equation, $\mathrm{Log} M_w = 1.327 \mathrm{Log}[\eta] + 4.371$, is obtained by a linear regression. It means that M_w *versus* $[\eta]$ relationship is in good conformity with the following equation:

$$[\eta] = K M_w^{\alpha} \tag{1}$$

Based on the calculation, the equation representing M_w *versus* $[\eta]$ relationship for KGM can be established as follows:

$$[\eta] = 5.06 \times 10^{-4} M_w^{0.754} \tag{2}$$

N. Kishida et al.[17] studied M_w *versus* $[\eta]$ relationship for partial methylated KGM and obtained the equation

$$[\eta] = 6.37 \times 10^{-4} M_w^{0.74} \tag{3}$$

In Eqs. (2) and (3) there are a close α but different K, which is related to the methylation modification of KGM. Eq. (2) can be conveniently used to estimate the molecular weight of KGM by a simple viscometric method.

By using Eq. (2), the [η] *versus t* relationship in Figure 1 and Figure 2 can be transformed into the M_w *versus t* relationship as shown in Figure 4 and Figure 5, respectively.

The above-described change in M_w and MWD suggests that the enzymatic degradation of KGM by β-mannanase proceeds in a random or an endowise mechanism, which is compatible with the catalytic feature of β-mannanase as an endo-acting enzyme.

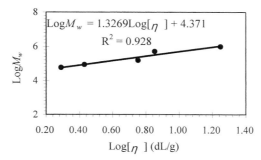

Figure 3. The plot of $LogM_w$ *versus* Log[η] for the aqueous solution of KGM samples prepared by enzyme-catalyzed degradation.

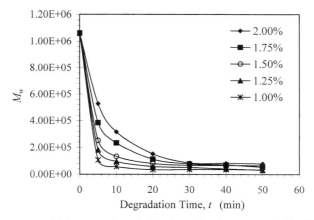

Figure 4. The curves of M_w *versus* degradation time t for the enzymatically degraded KGM samples by Mannanase I at different substrate concentrations.

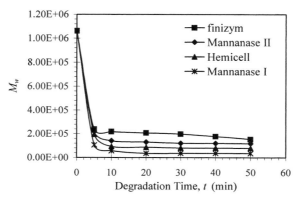

Figure 5. The curves of M_w *versus* degradation time t for the enzymatically degraded KGM samples by deferent enzymes at the same substrate concentration, 1.00 %.

Kinetics of Enzymatic Degradation. A first-order kinetics is generally used to describe the process of polymer degradation. It is expressed by

$$\frac{dL}{dt} = -kL \tag{4}$$

where L represents the total number of breakadable or hydrolyzable linkages in the system, k the apparent rate constant and dL/dt the reduction rate of L. Since there are $(M/m-1)$ linkages in each polymer molecule, where M is the average molecular weight of the macromolecular chain and m is the molecular weight of a monomeric unit. Let N_0 stand for the total number of polymer molecules, then L can be expressed as

$$L = N_0(M/m - 1) \tag{5}$$

If a polymer chain is long enough so as to $m/M \ll 1$, the following relationship between M and t can be derived

$$\frac{1}{M_t} = \frac{1}{M_0} + \frac{kt}{m} \tag{6}$$

in which M_0 and M_t are the molecular weight of the polymer at time $t = 0$ and t, respectively. Eq. (6) reveals the linear relation between the inverse M_t and degradation time t.

Therefore, we transformed the plots of the M_w *versus* t relationship in Figure 4 and Figure 5 into the plots of the inverse M_w *versus* t relationship as shown in Figure 6 and Figure 7, respectively. Obviously, regardless of the fact that in each degradation reaction the biocatalyst used or the substrate concentration is different, the plots in Figure 6 and Figure 7 all reveal a unique and two-stage linear relationship between the inverse M_w and degradation time t for the enzymatic degradation systems of KGM, which is consistent with Eq. (6) but exhibits two stages with different slopes. In the initial period of the degradation the straight lines of $1/M_w$ versus t exhibit greater slope than those obtained after the degradation has taken place to a certain extent.

Additionally, the slope of $1/M_w$ versus t, which is related to the apparent rate constant k, decreases with substrate concentration in the initial stage and tends to the same in the later/next stage. This kinetic behavior differs from the traditional kinetic model of polymer degradation in which no dependence of the rate constant on polymer concentration is considered. That is to say, the theory of a first-order reaction cannot explain the results shown in Figure 6.

The results obtained in our experiments is partly similar to those obtained by A. Tayal et al.[18] and Y. Cheng et al.[19] for the enzymatic degradation of water-soluble polysaccharide such as guar and its derivatives.

In order to make this problem clearer, assuming the enzymatic degradation reaction of KGM is following an nth-order kinetic process, then Eq. (4) becomes

$$\frac{dL}{dt} = -kL^n \quad (n \neq 1) \tag{7}$$

where n is the order of reaction. By integrating Eq. (7) between the limits $t = 0$ and t, and by doing appropriate substitutions, the following equation is obtained:

$$\left[\left(1 - \frac{m}{M_t}\right)^{1-n} - \left(1 - \frac{m}{M_0}\right)^{1-n}\right] = \frac{k(n-1)}{N_0^{1-n}} t \quad (n \neq 1) \tag{8}$$

For a zeroth-order reaction, $n = 0$, then Eq. (8) becomes

$$\frac{1}{M_t} = \frac{1}{M_0} + \frac{k}{mN_0} t \tag{9}$$

And for a second-order reaction, $n = 2$, then Eq. (8) becomes

$$\frac{1}{M_t} \cong \frac{1}{M_0} + \left(\frac{kN_0}{m}\right) t \tag{10}$$

Consequently the dependence of molecular weight on degradation time can be predicted by Eqs. (9), (6) and (10) for zero-, first- and second-order degradation reaction, respectively. Depending on the order of a reaction, the plot of $1/M_w$ versus t can be shown in the form of straight line whose slope is S. From Eqs. (9), (6) and (10) the slope of $1/M_w$ versus t plot for an nth-order reaction process can be generalized and expressed as follows:

$$S = \frac{kN_0^{n-1}}{m} \tag{11}$$

Let c_0 stand for the initial polymer or substrate concentration, V for the sample volume and N_{av} for the Avogadro number, then A. Tayal et al.[18] deduced the following equation:

$$c_0 = \frac{N_0 V}{N_{av} m} \tag{12}$$

By substitution we can get

$$S \propto c_0^{n-1} \tag{13}$$

It means that for a zero-order reaction ($n=0$), S or k is inversely proportional to c_0; and for a first-order reaction ($n=1$), S or k should be independent of c_0. This can be observed in the enzymatic degradation of KGM by analyzing the change in the slope of $1/M_w$ versus t plots. The result is shown in Figure 8. It suggests that the enzymatic degradation of KGM follows a zeroth-order reaction kinetics in the initial phase of the reaction and then a first-order reaction kinetics in the later phase over the whole range of substrate concentration studied.

We have noted the experiments on the enzymatic degradation of guar made by Y. Cheng etal.[19], which show that the substrate concentration range has a great influence on the kinetic behaviors of the enzymatic degradation of guar. At very low guar concentrations, the reaction rate increases with substrate concentration; at intermediate concentrations, the reaction rate becomes independent of substrate concentration; and at very high concentrations, the reaction rate decreases with substrate concentration. In view of the differences in the experiments done by us and by Y. Cheng et al., as well as the structural defferences between guar and KGM, it can be thought that our experimental results are compatible with those obtained by Y. Cheng et al. In fact, the 'real' substrate concentration, or breakable linkages, has been decreased in the later phase of the reaction. The theory concerning the formation of an enzyme-substrate complex and 'substrate saturation" can be used to explain these kinetic behaviors.   In the enzymatic degradation reactions with a given enzyme concentration, increasing the substrate concentration will certainly increase the ratio of substrate to enzyme. When the substrate concentration is high enough for the given enzyme concentration, each enzyme molecule can bind a molecule of substrate to form an enzyme-substrate complex.   In this case, increasing the substrate concentration cannot increase the number of complexes formed, and on the contrary increase the viscosity of a reaction system, thus leading to a diffusional resistance to the enzyme mobility in the concentrated polymer solution.   The reaction rate then becomes zero-order.

As for the fact that the plots of the inverse M_w versus t shown in Figure 7 also exhibit two-stage linear relationship with two distinct kinetic regimes, this indicates that the enzymatic degradation of KGM is a complex reaction where the kinetic mechanism is a combination of at least two reactions with different orders.   This may be a universal phenomenon for the degradation of polysaccharides.   The acid hydrolysis of cellulose[20] and carageenan [21, 22] also reflected the same kinetic behavior.

The above analysis indicates that the enzymatic degradation of KGM exhibits different kinetic behaviors under different conditions, and it is a complex reaction combining two or more reaction regimes with different orders.   The initial short-time reaction regime is of importance because a significant reduction in molecular weight of a substrate takes place in this reaction regime.   Therefore, the true order of reaction can be determined only by examining the effect of initial substrate concentration, c_0, on the apparent rate constant, k, or the slope of $1/M_w$ versus t plot, S.[18]

For the enzymatic degradation of a polysaccharide such as KGM, guar, cellulose etc, it is reasonable to be characterized using similar equations to Eqs. (9) and (6).   If it is assumed that an initial zeroth-order proceeds for time t' and is followed by a first-order reaction up to time t then, the following equations can be derived from Eqs. (9) and (6):

$$\frac{1}{M_{t'}} - \frac{1}{M_0} = \frac{k}{mN_0}t' \quad (0 < t < t')$$

(14)

$$\frac{1}{M_t} - \frac{1}{M'} = \frac{k_1}{m}(t - t') \quad (t' < t < t)$$

(15)

in which k and k_1 are the rate constants for the zeroth-order and the first-order reaction, respectively.

Eqs. (14) and (15) fit most of experimental data on the enzymatic degradation of polymers. Thus, this approach is of significance for objectively analyzing the kinetic behavior of polymer degradation previously ignored such as the degradation kinetics in the initial short-time phase.

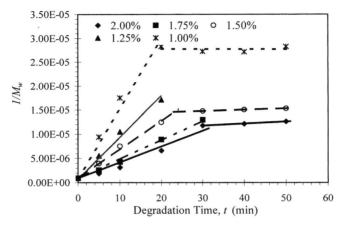

Figure 6. Reciprocal of molecular weight, $1/M_w$, as a function of degradation time t for the enzymatically degraded KGM samples by Mannanase I at various substrate concentrations.

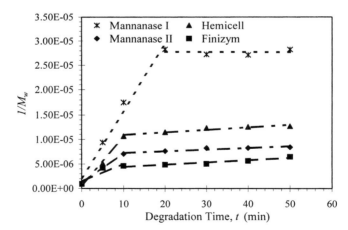

Figure 7. Reciprocal of molecular weight, $1/M_w$, as a function of degradation time t for the enzymatically degraded KGM samples by deferent enzymes at the same substrate concentration, 1.00 %.

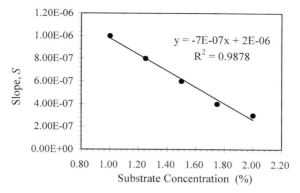

Figure 8. The slope of the straight line of $1/M_w$ versus c_0, S, as a function of KGM concentration in the initial period of enzymatic degradation.

Conclusion

1. The relation between $[\eta]$ and M_w of the KGM can be expressed by the equation $[\eta] = 5.06 \times 10^{-4} M_w^{0.754}$ or $\mathrm{Log}M_w = 1.327\mathrm{Log}[\eta] + 4.371$.

2. The enzymatic degradation of KGM catalyzed by Mannanase I is a complex reaction combining at least two reaction regimes with different orders over the range of substrate concentration (1.0 ~ 2.0 %) used in this work. In the initial phase (< 20 min) the degradation follows a zeroth-order reaction kinetics, i.e. its rate constant is inversely proportional to the initial substrate concentration; while in the next phase it follows a first-order reaction kinetics, i.e. the rate constant is independent of the initial substrate concentration. The theory concerning the formation of an enzyme-substrate complex and 'substrate saturation' have been used to explain this kinetic behavior.

3. The enzymatic degradation process of KGM catalyzed by other enzymes such as Mannanase II, Finizym and Hemicell is also a two-stage complex reaction with two distinct kinetic regimes.

These results are helpful to understand the kinetic behaviors of degradation of polysaccharides, thus achieving controllable or limited of KGM and other polysaccharides via an environmental benign enzyme-catalyzed reaction. However, more detailed investigation should be made.

178

Acknowledgements

The authors of this paper would like to acknowledge the 2001' Starting Grant from the Foundation of China Scholarship Council and the Foundation of Scientific and Technological Program of Guangzhou (Grant No. 2003J1-C0191) for the financial support.

[1]   Y. Nishio, T. Koide, Y. Miyashita, N. Kimura, *J. Polym. Sci., Part B: Polym. Phys.* **1999**, *37*, 1533.
[2]   K. Kato, K. Matsuda, *Agric. Biol. Chem.* **1969**, *33*, 1446.
[3]   C. Jia, S. Chen, W. Mo, Y. Meng, L. Yang, *Chinese Biochemical J.* **1988**, *4*, 407.
[4]   M. Maeda, H. Shimahara, N. Sugiyama, *Agric. Biol. Chem.* **1980**, *44*, 245.
[5]   FMC Corp., *Nutricol® Konjac General Technology Bulletin, Food Ingredients Division*, 1994.
[6]   V. Dave, M. Sheth, S. P. McCarthy, J. A. Ratto, D, L. Kaplan, *Polymer* **1998**, *39*, 1139.
[7]   M. Yoshimura, K. Nishinari, *Food Hydrocolloids* **1999**, *13*, 227.
[8]   Y. Sun, Q. Wu, G. Chen, X. Huang, *Food and Fermentation Industries (in Chinese)* **1999**, *25*, 47.
[9]   Y. Ohya, K. Ihara, J. Murata, T. Sugitou, T. Ouchi, *Carbohydrate Polymers* **1994**, *25*, 123.
[10]   Y. Zhang, X. Gan, Q. Zhou, L. Ma, *Pharmaceutical Biotechnology* **2001**, *8*, 200.
[11]   L. Qi, G. Li, M. Zong, *Acta Polymerica Sinica* **2003**, *5*, 650.
[12]   H. Shimahara, H. Suzuki, N. Sugiyama, K. Nisizawa, *Agric. Biol. Chem.* **1975**, *39*, 293.
[13]   H. Shimahara, H. Suzuki, N. Sugiyama, K. Nisizawa, *Agric. Biol. Chem.* **1975**, *39*, 301.
[14]   R. Takahashi, I. Kusakabe, S. Kusama, Y. Sakurai, K. Murakami, A. Maekawa, T. Suzuki, *Agric. Biol. Chem.* **1984**, *48*, 2943.
[15]   Z. He, J. Zhang, D. Huang, *Biotechnol. Lett.* **2001**, *23*, 389.
[16]   K. Kato, K. Matsuda, *Agric. Biol. Chem.* **1972**, *36*, 639.
[17]   N. Kishida, S. Okimasu, T. Kamata, *Agric. Biol. Chem.* **1978**, *42*, 1645.
[18]   A. Tayal, R. M. Kelly, S. A. Khan, *Macromolecules* **199**, *32*, 294.
[19]   Y. Cheng, R. K. Prud'homme, *Polym. Prep.* **2000**, *41*, 1868.
[20]   M. M. Figini, *J. Appl. Polym. Sci.* **1987**, *33*, 2097.
[21]   C. R. Masson, *Can. J. Chem.* **1955**, *33*, 597.
[22]   C. R. Masson, D. Santry, G. W. Caines, *Can. J. Chem.* **1955**, *33*, 1088.

Investigations of Miscibility in Interpenetrated Systems of Polyurethane and Polystyrene Obtained at Room Temperature

Jean-Michel Widmaier,* Jean-Marc Chenal

Institut Charles Sadron, CNRS - UPR 0022, 6, rue Boussingault 67083 Strasbourg, France
E-mail: jmw@ics.u-strasbg.fr

Summary: Interpenetrating polymer systems based on crosslinked polyurethane (PU) and polystyrene (PS) were prepared at room temperature by a one-shot (*in situ*) method, starting from an initial homogeneous mixture of reagents via non interfering mechanisms. Both polymerizations were performed either simultaneously or one after the other. Crosslinks and/or covalent bonds between components were deliberately introduced by the addition of appropriate monomers, in order to tailor the degree of microphase separation. Depending on the formation process, transluscent or transparent films were obtained, despite the difference in refractive index of the components. The maximum of miscibility, taken as from the glass transition criterion, was obtained for sequential tightly graft interpenetrating networks.

Keywords: dynamic mechanical analysis; *in situ* synthesis; interpenetrating polymer networks; phase separation

Introduction

Nowadays, most high performance materials are phase-separated materials resulting from the combination of existing polymers rather than issued from new monomers. The simplest type of combination is a mechanical blend of two polymers, polymer A and polymer B, which exhibits different properties, classically an elastomer associated with a rigid polymer. If possible from a synthetical point of view, formation of A-B block or graft copolymers is preferred in terms of their physical properties and phase-separated microstructure. But the newest multicomponent materials are interpenetrating polymer networks[1, 2], IPNs, which offer some synergy of properties due to their particular entangled morphology resulting from the control of the preparation conditions and composition. The so-called *in situ* IPNs are obtained by a one-shot process starting from an appropriate homogeneous mixture of monomers and prepolymers which

DOI: 10.1002/masy.200451217

is polymerized and crosslinked via non interfering reaction mechanisms, typically a step growth reaction and a chain polymerization. The onset of these two reactions may be concomitant or delayed in time, thus forming *in situ sim* (simultaneous) IPNs or *in situ seq* (sequential) IPNs, respectively. It is worth precising that for in situ *seq* IPNs, the onset of the second polymerization takes place well after gelation of the first component. Many kinds of interpenetrated systems can be formed, depending on the addition (or not) of multi- and/or hetero-functional molecules to the initial reactive mixture. They are denoted f-IPN (full) when both components are crosslinked, and s-IPN (semi) when only one component is crosslinked. IPN-G and IPN-g denote interpenetrated systems having short or long covalent bonds between the components, respectively.

The system under investigation is based on the combination of 35% polyurethane (PU) and 65% polystyrene (PS) by weight, which are two polymers thermodynamically incompatible. It is well known that microphase separation takes place in the course of the chemical reactions, according to a spinodal mechanism [3]. Incompatibility develops from a given conversion degree up to the formation of a sufficient number of entanglements which suppresses molecular mobility. Therefore, the conditions of formation reflect on the state of mixing. Various materials have thus been prepared, just differing by the presence or the absence of crosslinks and/or grafts. Information on the degree of phase separation was mainly gained from the glass transition behavior, determined by means of dynamic mechanical thermal analysis. Additional morphological informations were expected from small-angle X-ray scattering experiments.

Experimental

Interpenetrated PU/PS systems were prepared at room temperature by a one-shot process (*in situ* synthesis). Under vigorous stirring, a proper amount of poly(propylene oxide), MW~2000 g/mol, tris(6-isocyanato hexyl) isocyanurate, styrene and benzoin (0.5% by weight) were mixed for 5 min to form a homogeneous mixture. In some cases, 5 wt-% of divinylbenzene (DVB), and/or 1.5 wt-% of 2-hydroxyethyl methacrylate (HEMA) or 2-isocyanatoethyl methacrylate (IEM) was added to that mixture. The ratio between total isocyanate groups and total hydroxyl groups was 1.07. [4] Then, 1.5% by weight based on the PU components of dibutyltin dilaurate was added under stirring, and the mixture poured into a mold formed by two glass plates separated by an appropriate spacer and clamped together. Entrapped air bubbles were removed under vacuum.

The mold was exposed to the ultraviolet (UV) radiations (predominantly 365 nm) of a 100 W mercury lamp, either immediately after filling or after a 24 h preliminary stay in the dark. Light intensity at the surface was 7 mW·cm^{-2}. Reaction temperature was maintained at 25°C. After 10 h of irradiation which correspond to a plateau in the conversion versus time curves, light was switched off, and the sample was removed from the mold. Unreacted chemicals were eliminated under vacuum overnight. Films of 700 µm thickness were thus obtained.

A Fourier transform infra red spectrophotometer (Bomem Michelson MB 155) was used for kinetic measurements. The apparatus was equipped with an external UV light source (UVP SpotCure) and a flexible UV light guide, allowing simultaneous UV exposure and IR analysis. Scanning resolution was 2 cm^{-1} and 10 consecutive scans were averaged for each specimen. Reaction conversion was calculated from the decay of the normalized absorbance of characteristic peaks: the isocyanate peak at 2275 cm^{-1} and the vinyl double bond peak at 1639 cm^{-1}. Dynamic mechanical analysis (Metravib RAC 815 viscoanalyser) was performed to determine the glass transition temperature (Tg) of the films. Tg was taken at the maximum of the loss tangent (tan δ). Experiments in the tensile mode were done from –70 to 150°C at a fixed frequency of 5 Hz and a heating rate of 2°C/min, under dry nitrogen. The refractive index, n, of some samples was measured with an Abbe digital refractometer equipped with a thermostated (25°C) water circulation. Small-angle X-ray scattering (SAXS) experiments were carried out on a Nano Star apparatus using CuKα radiation. The signal was detected by a two-dimensional position-sensitive detector on dry films under vacuum. X-ray patterns were recorded every 10 min.

Results and Discussion

Two series of 35/65 PU/PS s-IPN, s-IPN-g (addition of IEM), s-IPN-G (addition of HEMA), f-IPN, f-IPN-g and f-IPN-G have been prepared, according to the above description. The first series concerns *in situ seq* materials whereas the second series refers to *in situ sim* materials. The reaction kinetics of both processes are illustrated in Figures 1 and 2 by their conversion versus time curves. Figure 1 shows that although both reactions start at the same time, the formation of PU and PS is not simultaneous. The PU formation appears to be faster than the polymerization of styrene, but gelation, which takes place around 70% conversion, occurs too late to prevent polymerization-induced phase separation to extend. On the other hand, concerning the *in situ seq*

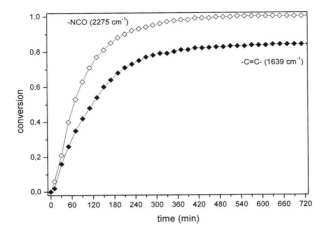

Figure 1. Kinetics of *in situ sim* f-IPN formation. PU ($\diamond$) ; PS ($\blacklozenge$).

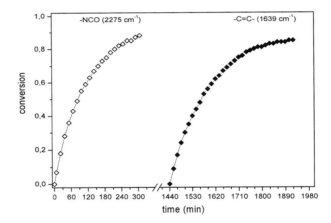

Figure 2. Kinetics of *in situ seq* f-IPN formation. PU ($\diamond$) ; PS ($\blacklozenge$).

process (Figure 2), the formation of the PU network is complete, and polymerization of styrene can only proceed in the immediate vicinity of the preformed PU network, due to topological hindrances. Therefore, higher miscibility between PU and PS is expected by using this sequential method.

Contrary to previous *in situ seq* full- and graft- IPNs based on the same precursor components and prepared similarly [5] (except for the use of a thermal initiator, which decomposes at 60°C, instead of a photoinitiator) which were opaque, the present *in situ sim* and *in situ seq* PU/PS films are slightly turbid and perfectly transparent, respectively (Figure 3).

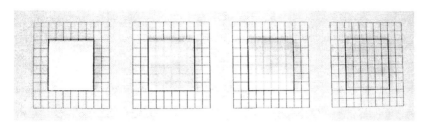

Figure 3. Aspect of 35/65 PU/PS f-IPN films obtained under various experimental conditions: from left to right: *in situ sim* (thermal decomposition of initiator), *in situ seq* (polymerization of styrene at 60°C), *in situ sim* (photopolymerization at room temperature), *in situ seq* (idem).

Since there is a significant difference in the refractive index of PU (n = 1.46) and PS (n = 1.59), the change in optical properties is related to the domain size, and transmission over 90% means that particle sizes under the wavelength of light are obtained when the process develops at room temperature due to higher viscosity which restrains complete phase separation. The refractive index of the films was measured (Table 1): except for the *sim* s-IPN, the refractive index of all samples is higher than the calculated value (n ~ 1.548) obtained by rule of mixture from the refractive index of the individual constituants. Systematically, the *in situ seq* film has a higher refractive index than the *in situ sim* film. Also, deviation from the calculated value increases when going from semi- to full-IPNs, and from loosely graft to tightly graft materials. Taking into account the Lorentz-Lorenz equation [6] which relates density and refractive index, the interpenetrated materials are more dense than the corresponding blends. This makes evident the presence of entanglements leading to forced miscibility which increases according to the previously mentioned order.

Table 1. Refractive index, n, of interpenetrated materials and deviation from simple additivity.

35/65 PU/PS	simple additivity	*in situ sim*	*in situ seq*	Deviation
s-IPN	1.5472	1.5478	1.5511	0.0039
f-IPN	1.5485	1.5517	1.5525	0.0040
s-IPN-g	1.5469	1.5527	1.5530	0.0061
f-IPN-g	1.5482	1.5538	1.5542	0.0060
s-IPN-G	1.5479	1.5544	1.5552	0.0073
f-IPN-G	1.5492	1.5560	1.5559	0.0069

$n_{PU} = 1.4647$; $n_{PU+IEM} = 1.4638$; $n_{PU+HEMA} = 1.4668$; $n_{PS} = 1.5916$; $n_{PS-co-DVB} = 1.5936$. Accuracy ± 0.0001.

The morphology of the various samples was investigated by using small-angle X-ray scattering. When plotting the scattering intensity versus the scattering vector, q, the SAXS curves of the

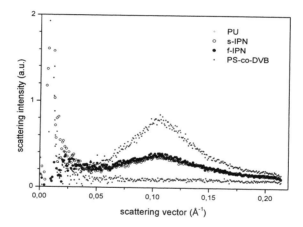

Figure 4. Small-angle X-ray scattering (SAXS) curves obtained for 35/65 PU/PS semi- and full-IPNs, compared to those of the individual components. *In situ seq* process.

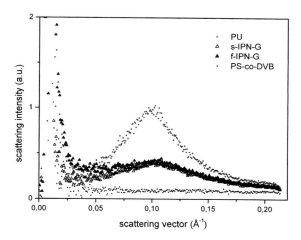

Figure 5. SAXS curves obtained for 35/65 PU/PS tightly graft semi- and full-IPNs. *In situ seq* process.

interpenetrated materials exhibit a broad maximum around 1.1 nm^{-1} , very similar in shape to the curve of the individual PU, but less intense, due to a compositional effect. The *in situ sim* and *in situ seq* curves are superimposable, but the curves of semi- and full-IPNs (Figure 4), and graft materials (Figure 5) are different, with significant broadening of the peak and increase of intensity towards smaller q.

This indicates less regular morphologies, i.e. enhanced miscibility. The peaks are too broad to determine the shift of the maximum with enough precision, however, a vague increase (from 5.98 nm to 6.16 nm) suggests that the PU phase is no longer a pure phase, but is enriched by some PS chains. Otherwise, the average distance between urethane groups would remain constant. Hence, these SAXS experiments are qualitative, only.

It is well-known that when a polymer mixture is phase-separated, two or more glass transitions, located at the temperature, T_g, of the individual components, are observed. Any shift in T_g accounts for the degree of miscibility. In general, the T_g of crosslinked polymers is difficult to detect using the classical DSC technique. Therefore, the more sensitive dynamic mechanical thermal analysis was used to determine the T_g of each component, which was taken as the temperature of the maximum of the loss factor, tan δ. Additionally, the half-peak width of tan δ is a good indication of the miscibility.

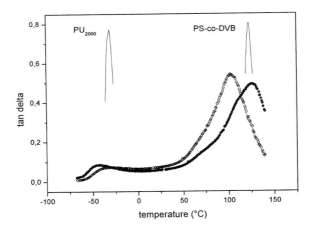

Figure 6. Temperature dependence of the loss tangent of 35/65 PU/PS full-IPNs formed by the *in situ sim* (●) or the *in situ seq* (○) process. The position of Tg of the individual PU and PS is indicated for comparison.

The two series of interpenetrated materials show two broad transitions, typical of phase-separated systems (Figure 6). Compared to the Tg of the individual PU and PS, there is no displacement of the transitions of the *in situ sim* films, but a significant inward shift is observed for the *in situ seq* films. For the latter, at low temperature, the transition of the PU-rich phase is shifted 8 to 23°C to higher temperatures, the maximum shift corresponding to the f-IPN-G sample. At the same time, the lower relaxation becomes less and less distinguishable (Figure 7). Similarly, the Tg of the PS-rich phase is shifted to lower temperatures: the lowest (11°C) and highest (37°C) shifts were for the s-IPN and f-IPN-G samples, respectively. Also, the magnitude of the tan δ peak, especially that of the PS-rich phase, increases together with some broadening of that transition for the *in situ seq* graft samples, compared to s-IPNs and f-IPNs. Consequently, the value of the loss factor in the intermediary region, i.e. between the lower and upper transitions, increases, thus rendering the *in situ seq* graft IPNs interesting as damping materials.

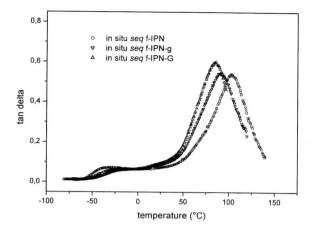

Figure 7. Temperature dependence of loss tangent of 35/65 PU/PS *in situ seq* materials: full-IPN (○), loosely graft IPN (▽) and tightly graft IPN (△).

Conclusion

Polyurethane and polystyrene are two very incompatible polymers. The introduction of steric hindrances like entanglements, crosslinks, covalent bonds increases the miscibility between polyurethane and polystyrene. Also, the *in situ* sequential synthesis of interpenetrated polymers, performed at room temperature and involving photopolymerization of styrene is more likely than the thermal *in situ* simultaneous process, from a miscibility point of view.

[1]   L. H. Sperling, "*Interpenetrating Polymer Networks and Related Materials*", Plenum Press, New York **1981**.
[2]   L. H. Sperling, V. Mishra, in: "*IPNs Around the World Science and Engineering*", S. C. Kim, L. H. Sperling, Eds., J. Wiley & Sons, New York **1997**, p. 1.
[3]   Y. S. Lipatov, *Prog. Polym. Sci.* **2002**, *27*, 1721.
[4]   M. T. Tabka, J. M. Widmaier, G. C. Meyer, *Macromolecules* **1989**, *22*, 1826.
[5]   V. Nevissas, J. M. Widmaier, G. C. Meyer, *J. Appl. Polym. Sci.* **1988**, *36*, 1467.
[6]   J. C. Seferis, in: "*Polymer Handbook*", 4th ed., J. Brandrup, E. H. Immergut, E. A. Grulke, Eds., J. Wiley & Sons, New York **1999**, p. 571.

Macromol. Symp. **2004**, *216*, 189-194

Rheological Behavior of Spinning Dope of Multiwalled Carbon Nanotube/Polyacrylonitrile Composites

Biao Wang,[1,2] *Jianmei Li,*[1] *Huaping Wang,*[1] *Jianming Jiang,*[1] *Yunqi Liu*[2]

[1] College of Materials Science & Engineering, Donghua University, Shanghai 200051, China
E-mail: wbiao2000@mail.dhu.edu.cn
[2] Center for Molecular Science, Institute of Chemistry, Chinese Academy of Sciences, Beijing 100080, China

Summary: The spinning dope of Multiwalled carbon nanotube/polyacrylonitrile composites was prepared by *in-situ* polymerization in 50wt% NaSCN solution. The rheological behavior of the dope containing carbon nanotubes from 0.19wt% to 1.9wt% was investigated using rotational rheometry at 60^0C. The nanotubes have diameters between 30–50nm and lengths ranging from 6 to 10 μm. The apparent viscosity decreases with the addition of carbon nanotubes; this can be explained by the inhibition of carbon nanotubes during polymerization progress. The spinning dope containing various contents of carbon nanotubes exhibits a shear thickening effect. However, the dynamic rheological test shows there is a peak of complex viscosity with the increasing in frequency.

Keywords: carbon nanotubes; *in-situ* polymerization; nanocomposites; polyacrylonitrile; rheology

Introduction

Composites of carbon nanotubes(CNTs) in polymeric matrices have attracted considerable attention in the research and industrial communities due to their unique mechanical and electrical properties. CNT/polymer nanocomposites possess high stiffness, high strength, and good electrical conductivity at relatively low concentrations of CNT filler.[1-3] A key issue in producing CNT/polymer nanocomposites is the ability to control dispersion of the CNTs in polymeric matrices. This, however, hinges on an in-depth understanding the rheological behavior of the CNT/polymer composites, a topic that has not received much attention.

Recently, the melt rheological properties of different CNT/polymer composites have been studied. For example, Petra Potschke et al.[4] examined the melt rheological properties of multiwalled carbon nanotube(MWNT) filled polycarbonate nanocomposites formed by melt extrusion. To our

DOI: 10.1002/masy.200451218

knowledge, there are no reports on the solution rheological behavior of spinning dope of MWNT/polyacrylonitrile(PAN) composites. The purpose of this paper is to study the effects of MWNT's addition on the rheological properties of the spinning dope. And the probable structure changings with the increase of shear rate have also been discussed.

Experimental

The MWNT were prepared in our laboratory, and the experimental procedure was described in our previous works[5]. The prepared MWNT were ultrasonically purified in mixture acid of HNO_3/H_2SO_4 (3:1, v/v), then cleaned with deionized water, filtrated and dried. After that, the MWNT dispersed in 50wt%NaSCN solvent was added to mixture solution containing acylonitrile (ACN), methyl methacrylate (MMA), itaconic acid (ITA), 2,2-azobisisobutyronitrile(AIBN) and 50wt% NaSCN. The mixture was stirred and heated at 78^0C for one hour. The spinning dope was obtained after demonomerizing. The process of polymerization was listed in Table 1.

Table 1. Process of polymerization

ACN (wt%)	MMA (wt%)	ITA (wt%)	PH (-)	Temp. ($^\circ C$)	Time (hr)
89.5	10	0.5	4.5	78	1

The purified MWNT samples were examined by TEM (JEOL 100CX2). The rheological measurements were performed using rotational rheometry RS1 (Haake) at 60^0C(using a cone plate geometry). The molecular weights of the samples were examined with a Waters gel permeation chromatograph (GPC, model 1500) with Ultrastyragel columns ($7.8 \times 300mm$), using $DMF/NaNO_3$ as the eluate. The experiments were carried out at 50^0C, with a pumping rate of 1ml/min.

Results and Discussion

TEM studies show that the prepared MWNT samples have diameters ranging from 30 to 50nm and length between 6 and 10um. Figure1 shows the TEM image of purified MWNTs. As can be seen from Figure 1, the surface of the nanotube is very clean. Most of the impurity phases such as amorphous carbon and graphitic nanoparticles are removed.

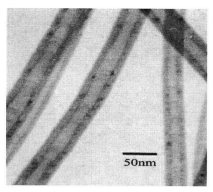

Figure 1. TEM image of purified MWNT.

The effect of MWNT contents on molecular weights and conversion rate is shown in Table 2. The number-average and weight-average molecular weights decrease with the increasing of MWNT content under the same polymerizing conditions. And it seems that there is little discernible trend observed between conversion rate and MWNT contents. These results indicate the inhibition of MWNT during polymerization progress. Because the nanotube can also be initiated by a free-radical initiator, AIBN, to open their π bonds[6].

Table 2. The effect of MWNT concents on molecular weights and conversion rate

Content of MWNT (wt%)	Number-average molecular weights	Weight-average molecular weights	Conversion rate (%)
0	6.74	12.01	53.95
0.19	6.57	11.73	53.56
0.53	6.33	10.31	53.93
0.94	6.21	10.07	50.50
1.90	5.85	9.54	54.68

Figure 2 shows the flow curves of spinning dope of the MWNT filled polyacrylonitrile(PAN) nanocomposites. From Figure 2, the apparent viscosity decreases with the increasing of MWNT contents. The inhibition of carbon nanotubes leads to lower the molecular weights of the dope, which makes the apparent viscosity decrease. When the contents of MWNT are more than 0.53wt%, the spinning dope exhibits a shear thickening effect. The measurements were also carried out in an oscillatory shear mode using a cone plate geometry(35mm diameter) at 60^{0}C. Figure 3 shows the complex viscosity of nanotube filled PAN spinning dope. For PAN sample, the complex viscosity decreases rapidly with the increasing of the frequency. However, for MWNT/PAN sample, the complex viscosity increases slightly with the frequency's increasing in lower frequency stage, and decreases in higher frequency stage. There is a peak of complex viscosity with the increasing in frequency. It is likely that the PAN segments near to the nanotube surface were less mobile and retarded orientation of PAN macromolecule, leading to the increasing in viscosity at lower shear rate. Similar effects have been observed in other polymer systems filled with finely divided materials[2]. Carbon nanotube, a linear molecular, have also oriented when the shear rate turn much higher. The orientation of MWNT and the PAN macromolecule make the viscosity decreased at higher shear rate.

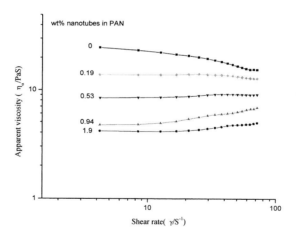

Figure 2. Flow curves of spinning dope of the MWNT filled polyacrylonitrile.

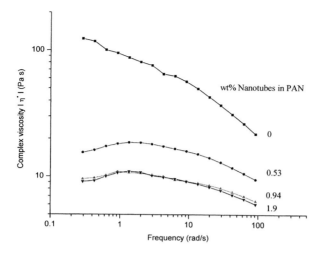

Figure 3. Complex viscosity of MWNT/PAN spinning dope.

Conclusion

The spinning dope of Multiwalled carbon nanotube/polyacrylonitrile composites were prepared with *in-situ* polymerization. The apparent viscosity decreases with the addition of carbon nanotubes. The spinning dope exhibits a shear thickening effect when the contents of carbon nanotubes are more than 0.53wt%. The dynamic rheological test shows that the complex viscosity increases slightly with the frequency's increasing in lower frequency stage, and decreases in higher frequency stage. The rheological behavior indicates that carbon nanotubes may retard the orientation of PAN molecules at lower shear stress and can oriented themselves in the dope at higher shear stress.

Acknowledgement

This project was supported by Science & Techology Committee of Shanghai and the Chinese Academy of Science.

194

[1] J. P. Salvetat, G. A. D. Briggs, J. M. Bonard *et al.*, *Phys. Rev. Lett.* **1999**, *82*, 944.
[2] M. S. P. Shaffer, A. H. Windle, *Adv. Mater.* **1999**, *11*, 937.
[3] D. Qian, E. C. Dickey, R. Andrews, T. Rantell, *Appl. Phys. Lett.* **2000**, *76*, 2868.
[4] P. Petra, T. D. Fornes, D. R. Paul, *Polymer* **2002**, *43*, 3247.
[5] W. B. Wang, Y. Q. Liu, D. B. Zhu, *Chem. Phys. Lett.* **2001**, *340*, 419.
[6] Z. Jia, Z. Wang, C. Xu, *et al.*, *Mater. Sci. and Eng. A.* **1999**, *271*, 395.

Macromol. Symp. **2004,** *216,* 195-208

Mathematical Analysis of the Formation of Molecule Sizes on a Spinning Disc Reactor

Katarina Novakovic, Julian Morris, Elaine Martin*

Centre for Process Analytics and Control Technology, School of Chemical Engineering and Advanced Materials, University of Newcastle, Newcastle upon Tyne, NE1 7RU, England
E-mail:e.b.martin@ncl.ac.uk

Summary: A mathematical analysis of the behaviour of the molecular weights of addition polymers during a polymerisation process is described. Spinning disc reactor (SDR) technology has been shown to yield significant improvements in terms of polymerisation rates whilst retaining close control of the molecular weights and the molecular weight distributions[1,2]. However, understanding of the kinetics of the polymerisation process on a SDR remains unresolved. One of the questions to be addressed concerns the sizes of the macromolecules preferably formed during the polymerisation process. To address this question, a mathematical analysis of the observed trends in number and weight average molecular weight, monomer concentration and polydispersity during the polymerisation process on a SDR has been undertaken. To validate the results, experimental data obtained from benzoyl peroxide initiated free radical polymerisation of styrene on a SDR[2] was used. It was concluded that most of the monomers consumed are in the growth of smaller size chains.

Keywords: addition polymerization; mathematical analysis; molecule sizes; polystyrene; spinning disc reactor

Introduction

The modelling of free-radical polymerisations is important both scientifically and industrially since the reaction mechanism is widely used for the synthesis of addition polymers. Spinning disc reactor (SDR) technology has been shown to yield significant improvements in polymerisation rates whilst maintaining tight control of the molecular weights and the molecular weight distributions[1,2]. The ability to utilise SDR technologies for free radical polymerisation would realise a new dimension in polymerisation processes. To achieve this goal, the determination of the kinetics of polymerisation processes on SDRs is a fundamental requirement. Consequently it

    DOI: 10.1002/masy.200451219

196

is important to understand the mechanisms of rotating disc reactors and be able to explain the observed phenomena, such as significant increases in conversion occurring with no increase in molecular weights. A study is presented whereby the intervals of the molecule sizes preferably formed on the SDR are determined for specific experimental conditions. The polymerization system[2] consisted of styrene as the monomer with an initial concentration of 7.28 mol/dm^3, benzoyl peroxide as initiator with an initial concentration of 5.1·10^{-2} mol/dm^3 and toluene as solvent with an initial concentration of 1.567 mol/dm^3. Polymerisation was started in a batch reactor with the product being feed to the SDR. In both reactors, the temperatures were initially set to 90°C. The results are presented in Figures 1 to 3.

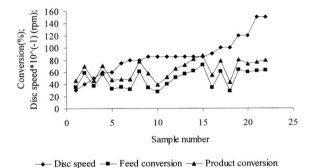

—◆— Disc speed —■— Feed conversion —▲— Product conversion

Figure1. Experimentally determined SDR feed and product conversion for different disc speeds.

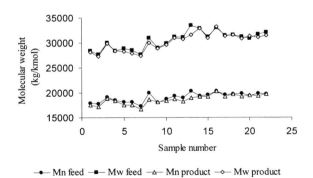

—●— Mn feed —■— Mw feed —△— Mn product —◇— Mw product

Figure 2. Molecular weights measured in the SDR feed and product.

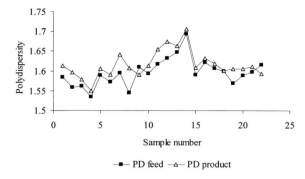

Figure 3. Polydispersity of SDR feed and product.

From the presented experimental data, a number of trends were identified. A slight decrease in the number average and weight average molecular weight of the SDR product was observed compared with the batch pre-polymerised SDR feed as well as a significant decrease in monomer concentration (i.e. increase in conversion) combined with a slight increase in polydispersity. These trends were mathematically analysed and solved independently. The solution of the individual analyses was a set of chains of specific length with the overall solution being the intersection of the individual solutions.

Mathematical Analysis of Trends Identified from Experimental SDR Data

Number average molecular weight is defined[3] as the total weight of all the molecules in a polymer divided by the total number of moles present:

$$M_n = \frac{\sum N_i \cdot M_i}{\sum N_i} \tag{1}$$

where the summation is over the different sizes of polymer molecules. M_n is the number average molecular weight, M_i is the weight of polymer molecule consisting of i monomer units and N_i is the number of moles whose weight is M_i. In the case of addition polymerisation, M_i can be defined as:

$$M_i = i \cdot M_{monomer} \tag{2}$$

where $i = (1,...,x)$ is the number of monomer units built in polymer molecule and $M_{monomer}$ is the molecular weight of the monomer unit. Combining Equations (1) and (2) gives:

$$\frac{\sum i \cdot N_i}{\sum N_i} = \frac{M_n}{M_{monomer}} \qquad (3)$$

Equation (3) can then be written as:

$$\frac{1 \cdot N_1 + 2 \cdot N_2 + ... + i \cdot N_i + ... + x \cdot N_x}{N_1 + N_2 + ... + N_i + ... + N_x} = \frac{(M_n)_{feed}}{M_{monomer}} \qquad (4)$$

where $(M_n)_{feed}$ is the number average molecular weight of SDR feed. From the experimental SDR data[2] presented in Figures 1 and 2, a decrease in the number average molecular weight was observed with the increase in conversion being significant. The trend of conversion is important for allowing differentiation between polymerisation and degradation. For example where the molecular weights are decreasing, it could be hypothesised that polymer degradation is occurring. However, for the case of a SDR reactor, it is evident from the data that an increase in conversion is present and therefore polymer degradation is not taking place. This observed behaviour can be included in Equation (4):

$$\frac{1 \cdot N_1 + 2 \cdot N_2 + ... + i \cdot (N_i + \Delta_i) + ... + x \cdot N_x}{N_1 + N_2 + ... + (N_i + \Delta_i) + ... + N_x} = \frac{(M_n)_{product}}{M_{monomer}} \qquad (5)$$

where Δ_i is the increase in number of moles of macromolecule with i monomer units, i.e. $\Delta_i = 1,2,3,...$ and $(M_n)_{product}$ is the number average molecular weight of SDR product. To simplify subsequent equations, $\frac{(M_n)_{feed}}{M_{monomer}}$ is denoted by k_1 and $\frac{(M_n)_{product}}{M_{monomer}}$ by k_2. Rearranging Equation (5) gives:

$$\frac{1 \cdot N_1 + 2 \cdot N_2 + ... + i \cdot N_i + ... + x \cdot N_x + i \cdot \Delta_i}{N_1 + N_2 + ... + (N_i + \Delta_i) + ... + N_x} = k_2 \qquad (6)$$

Furthermore:

$$\frac{k_1 \cdot (N_1 + N_2 + ... + N_i + ... + N_x) + i \cdot \Delta_i + k_1 \cdot \Delta_i - k_1 \cdot \Delta_i}{N_1 + N_2 + ... + (N_i + \Delta_i) + ... + N_x} = k_2 \qquad (7)$$

Equation (7) can be then rewritten as:

$$\frac{k_1 \cdot (N_1 + N_2 + ... + N_i + ... + N_x + \Delta_i) + \Delta_i \cdot (i - k_1)}{N_1 + N_2 + ... + (N_i + \Delta_i) + ... + N_x} = k_2 \tag{8}$$

or:

$$k_1 + \frac{\Delta_i \cdot (i - k_1)}{N_1 + N_2 + ... + (N_i + \Delta_i) + ... + N_x} = k_2 \tag{9}$$

To define a decrease in number average molecular weight, the following condition requires to be satisfied:

$$(M_n)_{product} < (M_n)_{feed} \tag{10}$$

This can be represented as:

$$k_1 + \frac{\Delta_i \cdot (i - k_1)}{N_1 + N_2 + ... + (N_i + \Delta_i) + ... + N_x} < k_1 \tag{11}$$

Since $\Delta_i > 0$ and $N_1 + N_2 + ... + (N_i + \Delta_i) + ... + N_x > 0$, the solution of Equation (11) is:

$$i < k_1, \text{ or after substitution } i < \frac{(M_n)_{feed}}{M_{monomer}} \tag{12}$$

However if the trend observed indicated that the number average molecular weight increases with time, alongside an increase in conversion, the condition to be satisfied becomes:

$$i > \frac{(M_n)_{feed}}{M_{monomer}} \tag{13}$$

To maintain a constant number average molecular weight over time whilst increasing conversion, an increase in the number of chains is necessary hence the following condition requires to be satisfied:

$$i = \frac{(M_n)_{feed}}{M_{monomer}} \tag{14}$$

A second analysis considered weight average molecular weight (M_w)[3]:

$$M_w = \frac{\sum N_i \cdot M_i^2}{\sum N_i \cdot M_i} \tag{15}$$

Substituting Equation (2) into Equation (15) gives:

$$\frac{\sum i^2 \cdot N_i}{\sum i \cdot N_i} = \frac{M_w}{M_{monomer}} \tag{16}$$

In expanded form, Equation (16) can be presented as:

$$\frac{1^2 \cdot N_1 + 2^2 \cdot N_2 + ... + i^2 \cdot N_i + ... + x^2 \cdot N_x}{1 \cdot N_1 + 2 \cdot N_2 + ... + i \cdot N_i + ... + x \cdot N_x} = \frac{(M_w)_{feed}}{M_{monomer}} \tag{17}$$

where $(M_w)_{feed}$ is the weight average molecular weight of the SDR feed. From the experimental SDR data[2] presented in Figures 1 and 2, a decrease in weight average molecular weight was detected, alongside a significant increase in conversion. This behaviour can be incorporated within Equation (17):

$$\frac{1^2 \cdot N_1 + 2^2 \cdot N_2 + ... + i^2 \cdot (N_i + \Delta_i) + ... + x^2 \cdot N_x}{1 \cdot N_1 + 2 \cdot N_2 + ... + i \cdot (N_i + \Delta_i) + ... + x \cdot N_x} = \frac{(M_w)_{product}}{M_{monomer}} \tag{18}$$

where $(M_w)_{product}$ is the weight average molecular weight of SDR product. Again, to simplify the subsequent equations, $\dfrac{(M_w)_{feed}}{M_{monomer}}$ is denoted as k_3 and $\dfrac{(M_w)_{product}}{M_{monomer}}$ as k_4. Rearranging Equation (18) gives:

$$\frac{1^2 \cdot N_1 + 2^2 \cdot N_2 + ... + i^2 \cdot N_i + ... + x^2 \cdot N_x + i^2 \Delta_i}{1 \cdot N_1 + 2 \cdot N_2 + ... + i \cdot (N_i + \Delta_i) + ... + x \cdot N_x} = k_4 \tag{19}$$

Furthermore:

$$\frac{k_3 \cdot (1 \cdot N_1 + 2 \cdot N_2 + ... + i \cdot N_i + ... + x \cdot N_x) + i^2 \Delta_i + k_3 \cdot i \cdot \Delta_i - k_3 \cdot i \cdot \Delta_i}{1 \cdot N_1 + 2 \cdot N_2 + ... + i \cdot (N_i + \Delta_i) + ... + x \cdot N_x} = k_4 \tag{20}$$

After rearranging, Equation (20) becomes:

$$\frac{k_3(1 \cdot N_1 + 2 \cdot N_2 + ... + i \cdot N_i + ... + x \cdot N_x + i \cdot \Delta_i) + i \cdot \Delta_i \cdot (i - k_3)}{1 \cdot N_1 + 2 \cdot N_2 + ... + i \cdot (N_i + \Delta_i) + ... + x \cdot N_x} = k_4 \tag{21}$$

or:

$$k_3 + \frac{i \cdot \Delta_i \cdot (i - k_3)}{1 \cdot N_1 + 2 \cdot N_2 + ... + i \cdot (N_i + \Delta_i) + ... + x \cdot N_x} = k_4 \tag{22}$$

To represent a decrease in weight average molecular weight, the following condition must be satisfied:

$$(M_w)_{product} < (M_w)_{feed} \tag{23}$$

This can be represented as:

$$k_3 + \frac{i \cdot \Delta_i \cdot (i - k_3)}{1 \cdot N_1 + 2 \cdot N_2 + ... + i \cdot (N_i + \Delta_i) + ... + x \cdot N_x} < k_3 \tag{24}$$

Since $\Delta_i, i > 0$ and $N_1 + 2 \cdot N_2 + ... + i \cdot (N_i + \Delta_i) + ... + x \cdot N_x > 0$ the solution of Equation (24) is:

$$i < k_3, \text{ or after substitution } i < \frac{(M_w)_{feed}}{M_{monomer}} \tag{25}$$

Alternatively if the weight average molecular weight increases with time, the condition to be satisfied is:

$$i > \frac{(M_w)_{feed}}{M_{monomer}} \tag{26}$$

To achieve a constant weight average molecular weight over time and an increase in conversion, an increase in number of chains must be achieved hence the following condition must be satisfied:

$$i = \frac{(M_w)_{feed}}{M_{monomer}} \tag{27}$$

Now considering polydispersity (PD):

$$PD = \frac{M_w}{M_n} = \frac{\sum N_i \cdot M_i^2 \cdot \sum N_i}{(\sum N_i \cdot M_i)^2} \tag{28}$$

According to Equations (9) and (22), the polydispersity of the SDR product $((PD)_{product})$ is given by:

$$(PD)_{product} = \frac{k_3 + \dfrac{i \cdot \Delta_i \cdot (i - k_3)}{1 \cdot N_1 + 2 \cdot N_2 + ... + i \cdot (N_i + \Delta_i) + ... + x \cdot N_x}}{k_1 + \dfrac{\Delta_i \cdot (i - k_1)}{N_1 + N_2 + ... + (N_i + \Delta_i) + ... + N_x}} \tag{29}$$

From the experimental data[2] presented in Figure 3, a slight increase in polydispersity of the SDR product compared with the polydispersity of the disc feed was observed. This can be represented as:

$$\frac{k_3 - \dfrac{i \cdot \Delta_i \cdot (k_3 - i)}{1 \cdot N_1 + 2 \cdot N_2 + ... + i \cdot (N_i + \Delta_i) + ... + x \cdot N_x}}{k_1 - \dfrac{\Delta_i \cdot (k_1 - i)}{N_1 + N_2 + ... + (N_i + \Delta_i) + ... + N_x}} > \frac{k_3}{k_1} \tag{30}$$

Furthermore:

$$\frac{k_3 \cdot \left(1 - \dfrac{i \cdot \Delta_i \cdot (k_3 - i)}{k_3 \cdot (1 \cdot N_1 + 2 \cdot N_2 + ... + i \cdot (N_i + \Delta_i) + ... + x \cdot N_x)}\right)}{k_1 \cdot \left(1 - \dfrac{\Delta_i \cdot (k_1 - i)}{k_1 \cdot (N_1 + N_2 + ... + (N_i + \Delta_i) + ... + N_x)}\right)} > \frac{k_3}{k_1} \tag{31}$$

$$\frac{1 - \dfrac{i \cdot \Delta_i \cdot (k_3 - i)}{k_3 \cdot (1 \cdot N_1 + 2 \cdot N_2 + ... + i \cdot (N_i + \Delta_i) + ... + x \cdot N_x)}}{1 - \dfrac{\Delta_i \cdot (k_1 - i)}{k_1 \cdot (N_1 + N_2 + ... + (N_i + \Delta_i) + ... + N_x)}} > \frac{1}{1} \tag{32}$$

$$\left(1 - \frac{i \cdot \Delta_i \cdot (k_3 - i)}{k_3 \cdot (1 \cdot N_1 + 2 \cdot N_2 + ... + i \cdot (N_i + \Delta_i) + ... + x \cdot N_x)}\right) > \left(1 - \frac{\Delta_i \cdot (k_1 - i)}{k_1 \cdot (N_1 + N_2 + ... + (N_i + \Delta_i) + ... + N_x)}\right)$$

$$\tag{33}$$

Solution of Equation (33) then requires the following equation to be solved:

$$\frac{k_3}{k_1} > \frac{\dfrac{i \cdot \Delta_i \cdot (k_3 - i)}{1 \cdot N_1 + 2 \cdot N_2 + ... + i \cdot (N_i + \Delta_i) + ... + x \cdot N_x}}{\dfrac{\Delta_i \cdot (k_1 - i)}{N_1 + N_2 + ... + (N_i + \Delta_i) + ... + N_x}} \tag{34}$$

After rearranging Equation (34) and including the constraint $\Delta_i > 0$, Equation (34) can be rewritten as:

$$k_1 \cdot i^2 - (k_1 \cdot k_3 + k_3 \cdot k_2) \cdot i + k_1 \cdot k_3 \cdot k_2 > 0 \tag{35}$$

The solution of the Equation (35) which represents a quadratic trinomial depends on the sign of its discriminant (D). If:

$$(k_1 \cdot k_3 + k_3 \cdot k_2)^2 - 4 \cdot (k_1)^2 \cdot k_3 \cdot k_2 < 0 \tag{36}$$

Then for every i, Equation (35) has the same sign as the i^2 coefficient. Else if:

$$(k_1 \cdot k_3 + k_3 \cdot k_2)^2 - 4 \cdot (k_1)^2 \cdot k_3 \cdot k_2 \geq 0 \tag{37}$$

The interval in which the set of solutions lies will depend on the coefficient of i^2. For the experimental SDR data, Equation (37) is satisfied and the coefficient of i^2 is positive at all times, $k_1 > 0$ i.e. $\dfrac{(M_n)_{feed}}{M_{monomer}} > 0$. Therefore the quadratic trinomial presented in Equation (35) will be satisfied for all i outside the interval given by the solutions, $i_{1,2}$, of the trinomial represented as the quadratic equation:

$$i_{1,2} = \frac{(k_1 \cdot k_3 + k_3 \cdot k_2) \pm \sqrt{(k_1 \cdot k_3 + k_3 \cdot k_2)^2 - 4 \cdot (k_1)^2 \cdot k_3 \cdot k_2}}{2 \cdot k_1} \tag{38}$$

or after substitution:

$$i_{1,2} = \frac{\left(\dfrac{(M_n)_{feed}}{M_{monomer}} \cdot \dfrac{(M_w)_{feed}}{M_{monomer}} + \dfrac{(M_w)_{feed}}{M_{monomer}} \cdot \dfrac{(M_n)_{product}}{M_{monomer}}\right)}{2 \cdot \dfrac{(M_n)_{feed}}{M_{monomer}}} \pm$$

$$\pm \frac{\sqrt{\left(\dfrac{(M_n)_{feed}}{M_{monomer}} \cdot \dfrac{(M_w)_{feed}}{M_{monomer}} + \dfrac{(M_w)_{feed}}{M_{monomer}} \cdot \dfrac{(M_n)_{product}}{M_{monomer}}\right)^2 - 4 \cdot \left(\dfrac{(M_n)_{feed}}{M_{monomer}}\right)^2 \cdot \dfrac{(M_w)_{feed}}{M_{monomer}} \cdot \dfrac{(M_n)_{product}}{M_{monomer}}}}{2 \cdot \dfrac{(M_n)_{feed}}{M_{monomer}}}$$

$$\tag{39}$$

To increase polydispersity:

$$i = (0, i_1) \text{ or } i = (i_2, \infty) \text{ and } i_1 < i_2 \tag{40}$$

It should be noted that $i > 0$ for all i. Furthermore if polydispersity was observed to be decreasing over time, the condition to be satisfied becomes:

$$i = (i_1, i_2) \tag{41}$$

To ensure polydispersity is constant over time whilst increasing conversion, an increase in the number of chains would be required and hence the following condition needs to be satisfied:

$$i = i_1 \text{ or } i = i_2 \tag{42}$$

The overall solution of the system of trends analysed is the intersection of the individual solutions with the molecule sizes preferably formed on the spinning disc reactor being determined as the intersection of the solutions of equations (12), (25) and (40).

Results and Discussion of the Analysis

The reaction system analysed on a spinning disc reactor was benzoyl peroxide initiated free radical polymerisation of styrene[2]. The temperature of the SDR was controlled to within 84-91°C, and the rotational speed applied was between 300 and 1500rpm. The SDR feed was pre-polymerised in a batch rector. Pre-polymerisation time was between 40 and 120 minutes. Data from twenty two runs on the SDR was made available. Two to four measurements were acquired for each run and the average value was considered. These results were obtained through standard laboratory experimentation procedures and did not utilise design of experiments.

For every set of data, the interval of chain lengths preferably formed was calculated. For a few runs the observed trend differed to that for the majority of runs. For example for some runs a slight decrease in polydispersity or slight increase in molecular weights was noted. These runs were also analysed and the results are presented. The monomer in system analysed was styrene:

$$M_{monomer} = M_{styrene} = 104 \text{ g/mol} \tag{43}$$

After undertaking the trend analysis for the twenty two experimental runs presented in Table 1, the majority (18 of the 22) lay within three intervals presented in Equation (44), (46) and (47). The largest group of experimental results (16 of the 22) fitted chain length interval of:

$$(0, i_1) \tag{44}$$

And with two of the individual runs (samples 9 and 22) lying within the following two intervals respectively:

$$\left(\frac{(M_n)_{feed}}{M_{monomer}}, \frac{(M_w)_{feed}}{M_{monomer}}\right) \tag{45}$$

$$\left(i_1, \frac{(M_n)_{feed}}{M_{monomer}}\right) \tag{46}$$

where:

$$i_1 = \frac{\left(\dfrac{(M_n)_{feed}}{M_{monomer}} \cdot \dfrac{(M_w)_{feed}}{M_{monomer}} + \dfrac{(M_w)_{feed}}{M_{monomer}} \cdot \dfrac{(M_n)_{product}}{M_{monomer}}\right)}{2 \cdot \dfrac{(M_n)_{feed}}{M_{monomer}}} -$$

$$- \frac{\sqrt{\left(\dfrac{(M_n)_{feed}}{M_{monomer}} \cdot \dfrac{(M_w)_{feed}}{M_{monomer}} + \dfrac{(M_w)_{feed}}{M_{monomer}} \cdot \dfrac{(M_n)_{product}}{M_{monomer}}\right)^2 - 4 \cdot \left(\dfrac{(M_n)_{feed}}{M_{monomer}}\right)^2 \cdot \dfrac{(M_w)_{feed}}{M_{monomer}} \cdot \dfrac{(M_n)_{product}}{M_{monomer}}}}{2 \cdot \dfrac{(M_n)_{feed}}{M_{monomer}}}$$

$$\tag{47}$$

Less than 5% of the data analysed (sample number 20) showed that chains that increased in number lay in the interval:

$$(i_2, \infty) \tag{48}$$

i.e. they favoured the production of bigger chains, where:

$$i_2 = \frac{\left(\dfrac{(M_n)_{feed}}{M_{monomer}} \cdot \dfrac{(M_w)_{feed}}{M_{monomer}} + \dfrac{(M_w)_{feed}}{M_{monomer}} \cdot \dfrac{(M_n)_{product}}{M_{monomer}}\right)}{2 \cdot \dfrac{(M_n)_{feed}}{M_{monomer}}} +$$

$$+ \frac{\sqrt{\left(\dfrac{(M_n)_{feed}}{M_{monomer}} \cdot \dfrac{(M_w)_{feed}}{M_{monomer}} + \dfrac{(M_w)_{feed}}{M_{monomer}} \cdot \dfrac{(M_n)_{product}}{M_{monomer}}\right)^2 - 4 \cdot \left(\dfrac{(M_n)_{feed}}{M_{monomer}}\right)^2 \cdot \dfrac{(M_w)_{feed}}{M_{monomer}} \cdot \dfrac{(M_n)_{product}}{M_{monomer}}}}{2 \cdot \dfrac{(M_n)_{feed}}{M_{monomer}}}$$

$$\tag{49}$$

These results are presented in Table 1 and Figure 4.

Table 1. Experimentally measured molecular weights and results from the analysis.

$S^{a)}$	$Mn^{b)}$ g/mol	$Mw^{b)}$ g/mol	$PD^{b)}$	$Mn^{c)}$ g/mol	$Mw^{c)}$ g/mol	$PD^{c)}$	D	i_1	i_2	Interval solution
1	17923	28402	1.585	17475	28192	1.613	3.19E+9	105.8	433.5	0,105
2	17732	27637	1.558	17118	27342	1.597	2.84E+9	104.8	417.5	0,105
3	19149	29888	1.561	18891	29824	1.579	3.97E+9	114.3	456.6	0,114
4	18529	28423	1.534	18318	28402	1.55	3.26E+9	111.4	432.1	0,111
5	18139	28814	1.589	17580	28220	1.605	3.36E+9	106.7	438.9	0,106
6	18106	28489	1.573	17544	27907	1.591	3.21E+9	106.8	432.5	0,106
7	17325	27631	1.595	16729	27465	1.642	2.82E+9	101.6	420.6	0,101
8	19993	30894	1.545	18647	29960	1.607	4.31E+9	116.3	457.8	0,116
9	17969	28923	1.61	18142	28861	1.591	3.53E+9	107.5	451.4	173,278
10	18703	29786	1.593	18365	29625	1.613	3.88E+9	110.7	456.9	0,110
11	19282	31185	1.617	18678	30907	1.655	4.57E+9	112.8	477.5	0,112
12	18997	31023	1.633	18307	30649	1.674	4.44E+9	110.5	475.3	0,110
13	20259	33392	1.648	18919	31486	1.664	5.76E+9	115.6	505.3	0,115
14	19338	32785	1.695	19246	32827	1.706	5.61E+9	113.1	515.9	$NS^{d)}$
15	19591	31165	1.591	19221	30905	1.608	4.65E+9	115.9	477.7	0,115
16	20332	32987	1.622	20300	33144	1.633	5.89E+9	120.6	513.2	$NS^{d)}$
17	19519	31359	1.607	19241	31275	1.619	4.77E+9	115.4	483.3	0,115
18	19698	31502	1.599	19675	31505	1.601	4.93E+9	117.4	488.0	$NS^{d)}$
19	19860	31168	1.569	19188	30823	1.606	4.6E+9	117.1	472.1	0,117
20	19353	30744	1.589	19446	31219	1.605	4.51E+9	115.9	476.7	477,∞
21	19816	31673	1.598	19291	31083	1.611	4.91E+9	116.6	484.4	0,116
22	19755	31933	1.616	19701	31385	1.593	5.17E+9	117.3	496.0	117,189

[a)] Sample number.
[b)] SDR feed.
[c)] SDR product.
[d)] No solution.

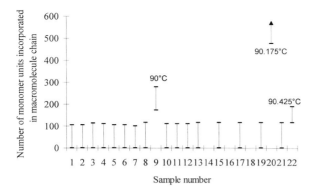

Figure 4. Calculated length of chains preferably increasing during the polymerisation process on the spinning disc reactor.

A small number of the experimental runs (samples 14, 16 and 18) had no mathematical solution as there was no intersection between the solutions of the individual analysis. After examining the data in more detail it was observed that the difference between the average molecular weight of product and feed was small and of opposite sign to that expected (i.e. for sample number 18 $(M_w)_{product}-(M_w)_{feed}=3$). This is most likely due to experimental error in the measurement of molecular weights. The difference in molecular weights could be of different sign in which case the results from the analysis would be feasible.

Conclusions

A mathematical methodology for the analysis of the trends for the addition polymer qualities of number average and weight average molecular weights and their calculated ratio, polydispersity, is proposed. The approach was developed for benzoyl peroxide initiated free radical polymerisation of styrene on a spinning disc reactor. From Table 1 and Figure 4 it was observed that for the majority of the experimental data, the results of the mathematical analysis for a polymerisation process performed on a spinning disc reactor, the number of smaller polymer chains increases. This is in agreement with the known theory of polymers and is confirmed mathematically. Further work is planned to determine the relationship between the variable(s) influencing the chain lengths preferably formed and the chain length itself since as it is possible to calculate chain length, this can be incorporated into the design and ultimately the control of the polymerisation process.

As a result of the number of runs being limited, it was not possible to draw any firm conclusions. However, one tentative observation that could be made with respect to the SDR temperature was that for those chains whose increase lay within a specific interval, or exceeded a specific value, the disc temperature was equal to, or above, 90°C whilst for the remaining runs it was below 90°C. The variables expected to have an effect on the products macromolecule chain length are disc rotation speed and SDR polymerisation time. For the data available no conclusive findings could be drawn and further experimental work is required.

Although the methodology is developed for polymerisation processes carried out on a spinning disc reactor (SDR), the approach is applicable to other reactor systems. Furthermore general rules can be determined so that for specific data, collected over time, on number and weight average molecular weight, for addition polymerisations undertaken in any rector system, intervals for polymer chain sizes preferably formed during the process can be calculated.

Acknowledgements

KN would like to thank the UK ORS Scheme and CPACT for providing funding for her PhD studies.

[1]   K. V. K Boodhoo, R. J. J. Jachuck, *Applied Thermal Engineering* **2000**, *20*, 1127.
[2]   K. V. K. Boodhoo, *"Spinning Disc Reactor for Polymerisation of Styrene"*, PhD Thesis, Chemical and Process Engineering, University of Newcastle, Newcastle upon Tyne **1999**.
[3]   G. G. Odian, *"Principles of Polymerisation"*, 3rd ed., John Wiley & Sons, New York **1991**.

Effect of Ground Rubber Powder on Properties of Natural Rubber

Shuyan Li, Johanna Lamminmäki, Kalle Hanhi*

Plastics & Elastomer Technology, Institute of Materials Science, P.O.Box 589, Tampere University of Technology, 33101, Finland
E-mail: kalle.hanhi@tut.fi

Summary: Ground rubber powder (GRP) with three different sizes was incorporated into nature rubber matrix with different loading. Cure characteristics, swelling behaviour, crosslink density, tensile fractured surface, and mechanical properties have been studied. Based on the cure characteristics, it is evident that the processability of the rubber compounds has not changed obviously with the different GRP loading. The introduction of GRP in virgin rubber leads to the increase in swelling degree and the decrease in crosslink density. Tensile strength, hardness and abrasion resistant deteriorate with the increase of GRP loading, but the tear resistance gets better. If the ground rubber particles are smaller, the properties are more similar to the virgin rubber. Because of the phase separation of the GRP and matrix, the properties get worse with the bigger ground rubber powder.

Keywords: fillers; mechanical properties; recycling; rubber; waste

Introduction

The environmental problems caused by waste rubbers and discarded tires are becoming more and more serious. Many attempts to recycle waste tires have been undertaken, for both environmental and economic reasons.[1-5] The recycling of the waste tire can be divided into four categories: energy generation, utilization of whole tires, reprocessing and utilization as ground rubber.[6, 7, 8] The materials incorporated with ground rubber powder and virgin rubber, or thermoplastics have many possible applications.[9] When the ground rubber powder is incorporated into virgin rubber compounds, the physical properties, especially the tensile strength, is decreased compared to the virgin rubber.[10] The introduction of ground rubber powder also causes the changes of cure behaviour and crosslink density, which are the results of the migration of sulphur or accelerator between virgin rubber matrix and the ground rubber

© 2004 International Union of Pure and Applied Chemistry

DOI: 10.1002/masy.200451220

vulcanizates.[11, 12] In this paper, the effects of different GRP sizes and loadings on these properties will be discussed.

Experimental

Materials

The virgin nature rubber used as a matrix in this study is SMR5 obtained from Teknikum Oy, Finland. Three different sizes of ground rubber powder (GRP), GRPA: 0-0.2mm, GRPB: 0.71-1mm, GRPC: 4mm, were purchased from Genan A/B, Demark (in Figure 1). The rubber powders were prepared from passenger car and light truck tires from which the free fibre and metal have been removed, the loading amounts were 10, 30, 50 phr respectively.

Figure 1. SEM of GRPB: 0.71-1mm .

Sample Preparation

The compound recipes are given in Table 1. The ground rubber powder was premixed with virgin rubber using an internal mixer at 60 °C, and then the ingredients except sulphur were incorporated. This was for better distribution and prevents the prevulcanization. The total mixing time was about 6 min. After which the premixture was softening in a two-roll mill at

60 °C and then sulphur was added, the mixing was continued for 2 minutes to get the sheet. The vulcanization was performed by compression moulding press at 160 °C. The vulcanization times, t_{90}, were determined using a linear Monsanto 100S rheometer.

Table 1. The compound recipes (phr).

Formulation (phr)	NR	GRP10	GRP30	GRP50
GRP(A,B,C)	0	10	30	50
SMR5	100	100	100	100
Zinc oxide	5	5	5	5
Stearic acid	2	2	2	2
N-220	35	35	35	35
CBS[a]	0.7	0.7	0.7	0.7
Sulphur	2.25	2.25	2.25	2.25

[a]:N-cyclohexyl-2-benzothiazyl sulphonamide.

Measurements

A Monsanto 100S oscillating disc rheometer was used to obtain the cure characteristics at temperature 160 °C according to ISO 3417. The tensile fractured surface of the compound was investigated with a scanning electron microscope (JEOL JSM–T100). Circular test pieces of diameter 10mm were cut from the vulcanized sheets (2 mm thickness), and soaked into toluene at room temperature (25 °C) until equilibrium to test the swelling degree. The crosslink density of the gels was calculated from Flory-Rehner equation.[13] The interaction parameter for the NR-toluene system is 0.39. Tensile strength and tear strength were performed with a Monsanto Tensometer 10 testing machine according to ISO 37 (Type 1) and ISO 34 (Type A). Abrasion resistance was carried out at Zwick with ISO 4649, an abrasive run of 40m, loading 10N. Hardness was tested based on ISO 7619 using a ShoreA durometer.

Results and Discussion

Cure Characterization and Morphology

The effect of GRP loading and particle size on the cure characteristics is shown in Figure 2 and Figure 3. As seen in Figure 2, the minimum torque (M_L) increased and maximum torque

(**M_H**) decreased on the rheometer curve with a higher loading amount of GRP. While, the scorch time (**t_{s2}**) and the optimum cure time (**t_{90}**) were slightly shorter when GRP was added to the NR matrix as shown in Figure 3.

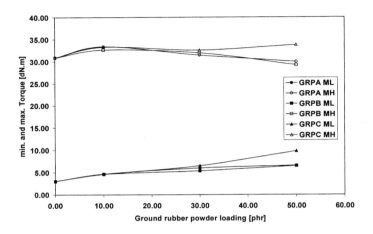

Figure 2. **M_L** and **M_H** *vs.* GRP loading.

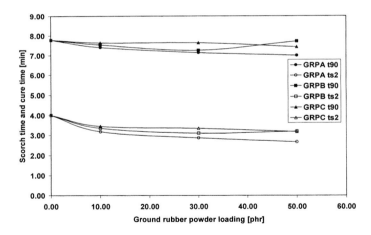

Figure 3. **t_{s2}** and **t_{90}** *vs.* GRP loading.

This is due to the migration of sulphur from the matrix rubber to the ground vulcanizates and the migration of accelerator fragments from the ground vulcanizates to the matrix.[14] However, at the same GRP loading, particle size does not affect the processability obviously. The morphology of the tensile fractured surface for GRPA 10 in Figure 4 shows the phase separation of GRP and matrix,[15] it is evident the deterioration of the mechanical properties for GRP introducing, particularly the tensile strength.

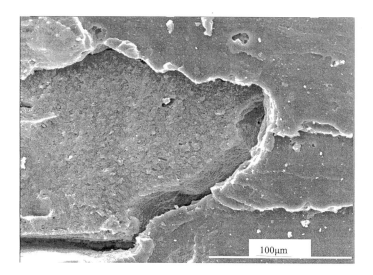

Figure 4. SEM of GRPA 10 tensile fractured surface.

Swelling Behaviour and Crosslink density

The swelling degree increases with the GRP loading in NR matrix as shown in Figure 5. The experiments indicate that the swelling degree increases with the bigger GRP particle size. This is attributed to the difficult diffusion of sulphur in the rubber matrix caused by ground vulcanizates.[12] The crosslink density also decreased with the GRP loading and bigger particle size, it is consistent with the result from cure characteristics.

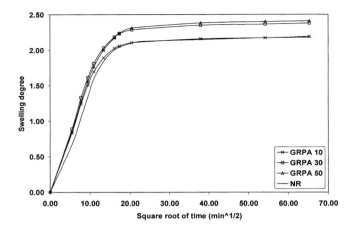

Figure 5. Swelling degree with GRPA loading.

Mechanical Properties

Tensile strength, tear strength, hardness and abrasion resistance are used to evaluate the mechanical properties of the system. The rubber compounds generally tend to become weak and brittle with as increasing loading volume and particle size. In spite of the general decrease in mechanical properties with increasing GRP loading, the properties retention values range from good to excellent at 10 phr loading with smaller particles especially with tensile strength and abrasion resistance as shown in Figure 6 and Figure 7.

Conclusion

It is evident that the processability of the system has not been changed obviously with the GRP loading from the cure characteristics. With the introducing of GRP in virgin rubber, it leads to the increase in swelling degree and the decrease in crosslink density. Tensile strength, hardness and abrasion resistant deteriorated with the increase of GRP loading. However, the tear resistance was improved. If the ground rubber particles are smaller, the properties are more similar to the virgin rubber. Because of the phase separation of the GRP and matrix, the properties got worse with the bigger ground rubber powder and higher loading.

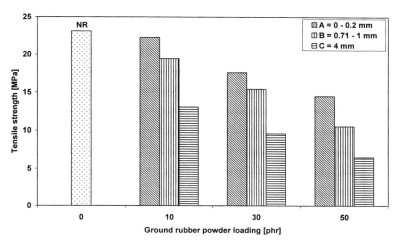

Figure 6. Tensile strength *vs.* GRP loading.

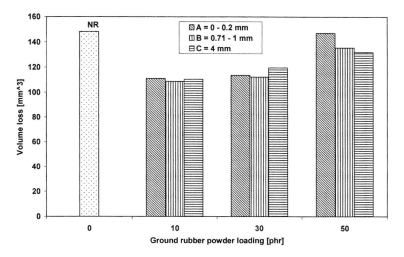

Figure 7. Volume loss *vs.* GRP loading.

216

[1]   K. Fukumori, M. Matsushita, H. Kamoto, *JSAE Review* **2002**, *23*, 259.
[2]   C. Jacob, A. K. Bhowmick, P. P. De, S. K. De, *Plastics, Rubber and Composites* **2002**, *31*, 212.
[3]   J. Yun, A. I. Isayev, *Polym. Eng. & Sci.* **2003**, *43*, 809.
[4]   O. Holst, B. Stenberg, M. Christiansson, *Biodegradation* **1998**, *9*, 301.
[5]   T. Luo, A. I. Isayev, *J. Elast. & Plast.* **1998**, *30*, 133.
[6]   B. Klingensmith, *Rubber World* **1991**, March, 16.
[7]   Y. Fang, M. Zhan, Y. Wang, *Materials & Design* **2001**, *22*, 123.
[8]   J. K. Kim, S. H. Lee, *J. Appl. Polym. Sci.* **2000**, *78*, 1573.
[9]   E. Kowalska, Z. Wielgosz, J. Pelka, *Polymers & Polymer Composites* **2002**, *10*, 83.
[10]   H. Ismail, R. Nordin, A. Md. Noor, *Polymer-Plastics Technology and Engineering* **2002**, *41*, 847.
[11]   C. Jacob, P. P. De, A. K. Bhowmick, S. K. De, *J. Appl. Polym. Sci.* **2001**, *82*, 3293.
[12]   H. Ismail, R. Nordin, A. M. Noor, *Polymer Testing* **2002**, *21*, 565.
[13]   M. Amin, G. M. Nasr, G. Attia, *Materials Letter* **1996**, *28*, 207.
[14]   S.-C. Han, M.-H. Han, *J. Appl. Polym. Sci.* **2002**, *85*, 2491.
[15]   T. Johnson, S. Thomas, *J. Mater. Sci.* **1999**, 34, 3221.

Preparation and Properties of Starch/Poly(vinyl alcohol) Composite Foams

Dujdao Preechawong, Manisara Peesan, Ratana Rujiravanit,* Pitt Supaphol*

The Petroleum and Petrochemical College, Chulalongkorn University, Bangkok 10330, Thailand
E-mail: ratana.r@chula.ac.th; pitt.s@chula.ac.th

Summary: Starch/poly(vinyl alcohol) (PVA) composite foams were prepared by baking a mixture of starch, PVA, and other ingredients in a hot mold. The effects of relative humidity, storage period, PVA content, and type and content of plasticizer (e.g. glycerol, urea, or ammonium chloride) on moisture content, water absorption, mechanical properties, and biodegradability of the as-prepared foams were investigated. In all of the testing conditions investigated, the ultimate tensile strength was shown to be a maximum when the relative humidity was 42%RH (for a fixed storage period of 7 days) and when the storage period was 2 days (for a fixed relative humidity of 42%RH). Addition of PVA improved the ultimate strength and the elongation at break of the composite foams. An increase in the amount of added plasticizer was responsible in an increase in the percentage of elongation at break at the expense of the tensile strength of the composite foams. Resistance to water absorption of starch/PVA composites foams was found to be superior to that of pure starch foams. Lastly, enzymatic degradation tests with α-amylase showed that addition of PVA affected a little to the enzymatic degradability of the composite foams.

Keywords: biodegradability; mechanical properties; moisture and water absorption; poly(vinyl alcohol); starch-based composite foam; starch foam

Introduction

Starch is a potentially useful material for making biodegradable plastics due mainly to its natural abundance, low cost, and thermoplastic properties.[1] Pure starch plastics, however, have some drawbacks, which include poor dimensional stability (a direct result of the high water absorption), poor mechanical properties, and poor processability.[2] To solve some of these problems, modification of starch molecules through various physical[3,4] or chemical[5] means have been considered.

Blending of starch or its derivatives, either in granular or destructurized form, with various polymers has been investigated. Starch has been added to a thermoplastic polymer, such as polyethylene (PE) and poly(vinyl chloride) (PVC), in order to impart biodegradable characteristics to the resulting blends. It has also been added to various expensive,

DOI: 10.1002/masy.200451221

biodegradable polyesters in order just to reduce the unit cost of the finished products. Some of the well-known examples are starch/polycarprolactone (PCL),[6,7] starch/poly(3-hydroxybutyrate-*co*-hydroxyvalerate),[6,8] and starch/poly(vinyl alcohol) (PVA) blends.[9-11]

It has been observed that addition of PVA helped improve moisture resistance to the baked foams.[9-11] Effects of added plasticizers have also been studied.[12-13] Under shearing and elevated temperature and pressure conditions, water and glycerol act as a good destructuring–plasticizing agent. Normally, plasticizers are added to adjust the properties of starch from being a soft (i.e. at high contents) to a brittle (i.e. at low contents) material.[14]

In the present contribution, biodegradable starch-based foams with good mechanical integrity have been prepared by blending tapioca starch with poly(vinyl alcohol) (PVA), along with other ingredients. Composite foams of various formulations have been prepared and analyzed for their mechanical properties, water absorption, and biodegradability.

Experimental Details

Materials

Tapioca starch was supplied as a courtesy from Siam Modified Starch Co., Ltd. (Thailand). Poly(vinyl alcohol) (PVA) having the degree of hydrolysis of around 97.5 to 99.5% was purchased from Fluka. Glycerol, urea, and ammonium chloride, used as plasticizers, were purchased from Carlo Erba. Guar gum was purchased from Sigma-Aldrich and magnesium stearate was supplied as a courtesy from Coin Chemical (Thailand) Co., Ltd. Both chemicals were used as additives. Termamyl 120 α-amylase (120 KNU/g) was supplied as a courtesy from East Asiatic Co., Ltd. (Thailand).

Preparation of Starch and Starch/PVA Foams

Tapioca starch (pre-dried at 110°C for 24 hours), guar gum (1% by weight of starch), and magnesium stearate (2% by weight of starch) were first dry-mixed using a Moulinex kitchen-aid mixer. Distilled water was then added to the mixture and the batter was further mixed for 20 min. For certain formulations, PVA or a plasticizer (i.e. glycerol, urea, or ammonium chloride) was also added to the batter. Starch and starch/PVA foams were then prepared by first applying 25 g of as-prepared batter in a picture-frame mold and then placing the mold in a Wabash V50H compression press (the temperature of the platens was pre-set at 220°C). After 2 min, the mold was cooled down to room temperature at a cooling rate that was fitted well by an exponential decay with a time constant of around 3 min.

Characterization

Morphology of starch and starch/PVA foams was examined using a JEOL JSM 520-2AE scanning electron microscope (SEM). The operating voltage used was 10 kV. Some fractured specimens obtained after mechanical property measurement were selected and were cut about 2 mm below the fractured surface and mounted on aluminum stubs, after which they were coated with a thin layer of gold.

Before moisture content determination and mechanical property measurement, starch and starch/PVA foams were conditioned in a conditioning jar having a specific relative humidity level of 11, 32, 42, 52, 67, or 75%RH at 25°C for various storage periods up to 7 days. The conditioning jars were prepared by filling the jars with saturate, aqueous solutions of LiCl, $MgCl_2$, K_2CO_3, $Mg(NO_3)_2$, $CuCl_2$, and NaCl, respectively.

The percentage of moisture content in a foam specimen (dried at 50°C for 24 hours) was taken as the percentage of weight increase after the specimen was conditioned in a specified relative humidity atmosphere for a specified storage period. The percentage of water absorption in a foam specimen (equilibrated at 50%RH for 7 days) was taken as the percentage of weight increase after the specimen was soaked in 100 ml of distilled water at 25°C for 10 min.

After the foam specimens were conditioned in a specified relative humidity atmosphere for a specified storage period, they were tested for various mechanical properties, e.g. tensile strength, percentage of elongation at break, flexural strength, and maximum flexural strain. These measurements were carried out using a Lloyd LRX mechanical testing machine, with the maximum load of 2500 N. For tensile tests, a crosshead speed of 5.2 mm/min and gauge length of 118 mm were used, while, for flexural tests, the crosshead speed of 1.3 mm/min and the span of 50 mm were applied.

Selected starch and starch/PVA foams were also tested for their enzymatic degradability. Each foam specimen was weighed before being placed in a test vial. A reaction mixture containing 25 ml of 0.05 M acetate buffer (pH 6.0), 1 ml of Termamyl α-amylase (120 KNU/g), and 54 mM of $CaCl_2 \cdot 2H_2O$ was added to the vial, which was then heated in a shaking water bath at 60°C for 3 hours. Specimens were collected after different reaction time periods, ranging from 5 to 180 min, were reached prior to being washed with distilled water and dried at 60°C for 5 hours. Their weights were then recorded and used to determine the percentage of weight loss.

Results and Discussion

Morphology

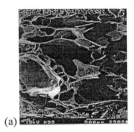 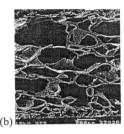

(a)   (b)

Figure 1. Scanning electron micrograph of cross sections of (a) starch and (b) starch/PVA (30 wt.%) composite foams.

Selected scanning electron micrographs of starch and starch/PVA foams are shown in Figure 1. Both micrographs show that the skin layers for both types of foam exhibited small, dense, and closed cell structure, whereas the interior showed large, loose, and opened cell structure. The densified outer skin layer was caused by abrupt evaporation of the batter layer close to the hot surface of the mold, causing the batter to dry very rapidly. The large, loose, and opened cell structure of the interior was a result of the large amount of steam venting out of the mold, causing cell walls to rupture.[10] Even though the skin layer of starch/PVA foam was similar to that of the starch foam, the interior structure was a bit denser. This is verified by the fact that the density of this starch/PVA foam (i.e. 0.225 g/cm^3) was higher than that of the pure starch foam (i.e. 0.138 g/cm^3).

Moisture and Water Absorption

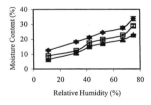 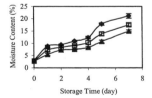

(a) (storage condition: 25°C for 7 days)   (b) (storage condition: 25°C and 42%RH)

Figure 2. Moisture content of starch and starch/PVA foams as a function of (a) relative humidity and (b) storage period. Keys: (♦) 0, (□) 10, and (▲) 30 wt.% PVA.

A number of factors (e.g. relative humidity, storage period, PVA content, and plasticizer content) affect moisture content of starch and starch/PVA foams. The effects of relative humidity, storage period, and PVA content on moisture content of various starch-based foams are illustrated in Figure 2. The moisture content in all of the foams studied (after being conditioned in a specified relative humidity atmosphere for 7 days at 25°C) was found to increase with increasing relative humidity and, for a given relative humidity level, it was found to decrease with increasing PVA content. For a fixed relative humidity level of 42%RH at 25°C, the moisture content in all of the foams investigated was found to increase monotonically with increasing storage period, and, for a given storage period, the moisture content was, again, found to decrease with increasing PVA content. The decrease in the tendency to absorb moisture with increasing PVA content is due to the fact that PVA is less hygroscopic than starch is.

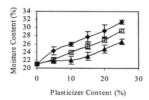

Figure 3. Moisture content of pure starch foams as a function of plasticizer content. The storage condition was at 42%RH and 25°C for 7 days. Keys: (◆) glycerol, (□) urea, and (▲) ammonium chloride.

The moisture content of pure starch foams containing glycerol, urea, or ammonium chloride after being conditioned at 42%RH and 25°C for 7 days is shown in Figure 3. Apparently, starch foams containing glycerol had a higher tendency to absorb moisture than those containing urea or ammonium chloride and, for a given type of added plasticizer, the moisture content was found to increase with an increase in its content.

Water absorption of starch and starch/PVA foams with varying PVA content has been investigated and it was found to decrease with increasing PVA content (i.e. from $127 \pm 4\%$ for pure starch foam down to $88 \pm 2\%$ and to $74 \pm 3\%$ for starch/PVA foams having the PVA content of 10 and 30 wt.%, respectively). It was postulated that a decrease in the water absorption for starch/PVA foams with increasing PVA content is due to the fact that a highly hydrolyzed PVA (for the particular PVA resin, the degree of hydrolysis was around 97.5 to 99.5%) is highly crystalline and insoluble in cold water.

Mechanical Properties

Effect of Relative Humidity

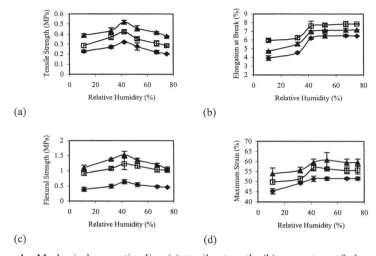

(a)   (b)

(c)   (d)

Figure 4. Mechanical properties [i.e. (a) tensile strength, (b) percentage of elongation at break, (c) flexural strength, and (d) maximum flexural strain] of starch and starch/PVA foams as a function of relative humidity. The storage condition was at 25°C for 7 days. Keys: (♦) 0, (□) 10, and (▲) 30 wt.% PVA.

Mechanical properties of starch-based foams (after being conditioned in a specified relative humidity atmosphere for 7 days at 25°C) were measured and the results are shown as a function of relative humidity in Figure 4. Apparently, all of the starch-based foams investigated exhibited a maximum in both tensile and flexural strength at the relative humidity level of 42%RH. At "low" relative humidity levels, the cell structure of starch-based foams was so brittle that cracks could form and propagate, whereas, at "high" relative humidity levels, the high amount of absorbed moisture weakened their cell structure.[11] It was also obvious that, for a given value of relative humidity, addition of PVA was responsible for a significant increase in both the tensile and flexural strength of the foams. Both percentage of elongation at break and maximum flexural strain of the foams were found to increase initially with increasing relative humidity level (up to 42%RH) and then became constant at relative humidity levels higher than 42%RH. An increase in the mobility of the starch molecules as a result of the absorbed moisture should be the reason for the observed increase in both the percentage of elongation at break and maximum flexural strain with increasing relative humidity level.[11]

Effect of Storage Period

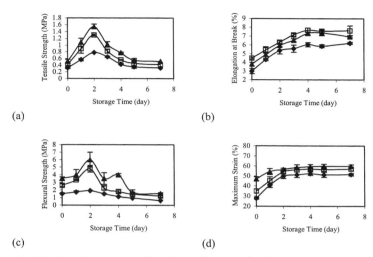

(a) (b)

(c) (d)

Figure 5. Mechanical properties [i.e. (a) tensile strength, (b) percentage of elongation at break, (c) flexural strength, and (d) maximum flexural strain] of starch and starch/PVA foams as a function of storage period at 42%RH and 25°C. Keys: (♦) 0, (□) 10, and (▲) 30 wt.% PVA.

The effect of storage period on mechanical properties of starch-based foams is illustrated in Figure 5. Obviously, all of the starch-based foams studied attained a maximum in both tensile and flexural strength after conditioning at 42%RH and 25°C for 2 days and the balance between the rigidity of the unplasticized or lightly-plasticized starch molecules and the mobility of the plasticized starch molecules may be responsible for the observed maximum in both of the strength values. It was also apparent that, for a given storage period, addition of PVA was responsible for an increase in both the tensile and flexural strength of the foams. Figure 5b shows that the percentage of elongation at break of the foams increased steadily with increasing storage period during the first four days of conditioning, after which it was found to vary very slightly. The rapid uptake of moisture during the first four days of conditioning may be responsible for the observed steady increase in the percentage of elongation at break of the foams.[15] The maximum flexural strain of the foams was also found to increase steadily with increasing storage period during around the first three to four days of conditioning, after which it was found to be practically unchanged.

Effect of PVA Content

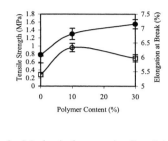   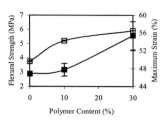

(a)   (b)

Figure 6. Mechanical properties [i.e. (●) tensile strength, (○) percentage of elongation at break, (■) flexural strength, and (□) maximum flexural strain] of starch and starch/PVA foams as a function of PVA content. The storage condition was 42%RH and 25°C for 2 days.

The effects of PVA addition and content on mechanical properties of starch-based foams after conditioning at 42%RH and 25°C for 2 days are shown graphically in Figure 6. Both tensile strength and flexural strength of the starch-based foams investigated were found to increase with increasing PVA content. Basically, PVA has higher tensile strength than starch (i.e. 95 versus 22 MPa); therefore, addition of PVA in the foam structure should help improve the tensile property as well as the rigidity of the resulting composite foams. The percentage of elongation at break of the foams was first found to increase with addition of 10 wt.% PVA and then decrease with further addition of PVA (up to 30 wt.%). The initial increase in the percentage of elongation at break for starch/PVA (10 wt.%) foam might be explained based on the higher flexibility of PVA versus starch molecules[10] and on the ease for crack formation at the interface between PVA and starch phases. Upon further increase in the PVA content, the rigidity of the foams increased appreciably (a direct result of the higher tensile strength of PVA versus starch),[16] leading to a reduction in the percentage of elongation at break for starch/PVA (30 wt.%). The maximum flexural strain of the foams was found to increase appreciably with addition of 10 wt.% PVA and then increase very slightly with further increase in the PVA content (up to 30 wt.%). The most likely explanation for such improvement in the maximum flexural strain may be due to the inherent flexibility of PVA molecules in comparison with starch molecules.

Effect of Plasticizer

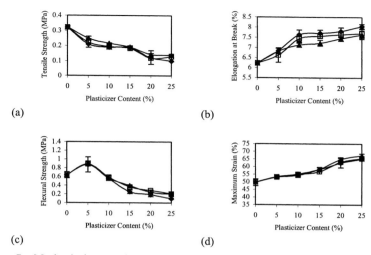

Figure 7. Mechanical properties [i.e. (a) tensile strength, (b) percentage of elongation at break, (c) flexural strength, and (d) maximum flexural strain] of pure starch foams as a function of plasticizer content. The storage condition was 42%RH and 25°C for 7 days. Keys: (♦) glycerol, (□) urea, and (▲) ammonium chloride.

The general purpose for incorporating a plasticizer into a plastic is to convert an otherwise hard and rigid plastic into a flexible or semi-flexible one. Since starch molecules are inherently rigid, addition of a plasticizer should make a starch-based article more pliable. In addition to moisture, three plasticizers (i.e. glycerol, urea, and ammonium chloride) are individually added into pure starch foams in order to study their effect on mechanical properties of the foams.

The effect of added plasticizer on mechanical properties of pure starch foams, after conditioning at 42%RH and 25°C for 7 days, is illustrated in Figure 7. As we normally expected, the tensile strength of the foams investigated was found to decrease, while the resulting percentage of elongation at break was found to increase, with addition and increasing amount of added plasticizer. This is probably because plasticizer molecules soften starch molecules by penetrating into starch granules and basically destroying the inter-molecular hydrogen bonds of starch molecules.[17] This, in turn, helps increase mobility of the starch molecules, resulting in decreased tensile strength and increased percentage of elongation at break. In case of flexural properties, the flexural strength of the foams was initially found to increase with increasing plasticizer content of up to 5 wt.%. Further increase in the plasticizer content greater than 5 wt.% resulted in the reduction of the flexural strength of the plasticized foams. The maximum flexural strain of the foams, however, was found to increase gradually with addition and increasing amount of added plasticizer. It is obvious that the type of plasticizer added did not, at all, affect the mechanical properties of the foams.

Enzymatic Degradation

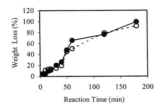

Figure 8. Enzymatic degradation by α-amylase at 60°C for (●) pure starch foams and (○) starch/PVA (30 wt.%) composite foams.

Enzymatic degradation of starch and starch/PVA foams was tested by α-amylase in vitro at 60°C and the results are plotted as a function of reaction time in Figure 8. All of the foams investigated exhibited a monotonic increase in the percentage of weight loss with increasing reaction time. An abrupt increase in the percentage of weight loss was observed at the reaction time of around 40 min. During "small" reaction times (i.e. less than 40 min), water molecules are being absorbed into the foam structure and this could be an explanation for the slow enzymatic degradability observed within this region. It is also apparent from Figure 8 that addition of PVA had little effect on the enzymatic degradability of the foams.

Conclusions

In the present contribution, starch/poly(vinyl alcohol) (PVA) composite foams were prepared by baking a mixture of starch, PVA, and other ingredients in a hot mold. The effects of conditioning relative humidity, conditioning storage period, PVA content, and type and content of added plasticizer (e.g. glycerol, urea, or ammonium chloride) on moisture and water absorption, mechanical properties, and biodegradability of the as-prepared composite foams were investigated and the results were compared with those of the pure starch foams.

Moisture absorption for both starch and starch/PVA foams was found to increase with increasing conditioning relative humidity level (the storage condition was 25°C for 7 days) and conditioning storage period (the storage condition was 42%RH and 25°C). Addition of PVA helped promote the resistance to water of the composite foams. In all of the testing conditions investigated, the ultimate tensile strength was shown to be a maximum when the relative humidity was 42%RH (for a fixed storage period of 7 days) and when the storage period was 2 days (for a fixed relative humidity of 42%RH). Addition of PVA improved the ultimate strength and the elongation at break of the composite foams. An increase in the amount of added plasticizer was responsible in an increase in the percentage of elongation at break at the expense of the tensile strength of pure starch foams. Lastly, the enzymatic degradability of both starch and starch/PVA foams by α-amylase was found to increase with increasing reaction time and addition of PVA affected a little to the enzymatic degradability of the composite foams.

Acknowledgments

This work is financially supported by Chulalongkorn University through a grant provided by the Rachadapisek Somphot Endowment Fund. The authors wish to thank Siam Modified Starch Co., Ltd. (Thailand), Coin Chemical (Thailand) Co., Ltd., and East Asiatic Co., Ltd. (Thailand) for supplying materials and essential chemicals for this work. They also due their gratitude to David C. Martin of the University of Michigan, USA, and John W. Ellis of LabTech Engineering Co., Ltd. (Thailand) for their technical knowledge and helpful suggestions.

[1]   W. M. Doane, *Starch - Stärke* **1992**, *44*, 293.
[2]   K. Tiefenbacher, *J. Macromol. Sci. Pure. Appl. Chem.* **1993**, *A30*, 727.
[3]   L. Averous, N. Fauconier, L. Moro, C. Fringant, *J Appl. Polym. Sci.* **2000**, *76*, 1117.
[4]   A. Roberta, I. Salvatore, N. Luigi, *J. Appl. Polym. Sci.* **1998**, *68*, 739.
[5]   A. D. Sugar, E. W. Merrill, *J. Appl. Polym. Sci.* **1995**, *58*, 1647.
[6]   M. F. Koening, S. J. Huang, *J. Polym. Mater. Sci. Eng.* **1992**, *67*, 290.
[7]   C. H. Kim, K. Y. Cho, J. K. Park, *Polym. Eng. Sci.* **2001**, *41*, 542.
[8]   M. A. Knotnis, G. S. O'Brien, J. L. Willett, *J. Env. Polym. Deg.* **1995**, *3*, 97.
[9]   R. L. Shogren, J. W. Lawton, *US Patent 5,756,194* (**1998**).
[10]   R. L. Shogren, J. W. Lawton, W. M. Doane, K. F. Tiefenbacher, *Polymer* **1998**, *39*, 6649.
[11]   R. L. Shogren, J. W. Lawton, L. Chen, K., F. Tiefenbacher, *J. Appl. Polym. Sci.* **1998**, *68*, 2129.
[12]   C. L. Swanson, R. L. Shogren, G. F. Fanta, S. H. Imam, *J. Env. Polym. Deg.* **1993**, *1*, 155.
[13]   D. Lourdin, L. Coignard, H. Bizot, P. Colonna, *Polymer* **1997**, *38*, 5401.
[14]   D. Lourdin, H. Bizot, P. Colonna, *J. Appl. Polym. Sci.* **1997**, *63*, 1047.
[15]   P. J. Stenhouse, J. A. Ratto, N. S. Schneider, *J. Appl. Polym. Sci.* **1997**, *64*, 2613.
[16]   L. Zhiqiang, F. Yi, Y. Xiao-Su, *J. Appl. Polym. Sci.* **1999**, *74*, 2667.
[17]   Y. Jingao, C. Songzhe, G. Jianping, Z. Huawu, Z. Jie, T. L. Tanjin, *Starch - Stärke* **1998**, *50*, 246.

Macromol. Symp. **2004**, *216*, 229-239

Preparation and Shelf-Life Stability of Aqueous Polyurethane Dispersion

Charoen Chinwanitcharoen,[1] *Shigeyoshi Kanoh,*[1] *Toshiro Yamada,**[1] *Kaoru Tada,*[1] *Shunichi Hayashi,*[2] *Shunji Sugano*[3]

[1] Department of Chemistry and Chemical Engineering, Kanazawa University, 2-40-20, Kodatsuno, Kanazawa, Japan
E-mail: tyamada@t.kanazawa-u.ac.jp
[2] Advanced Technology Research Center, Mitsubishi Heavy Industries, Japan
[3] Negami Chemical Industrial, Japan

Summary: A series of aqueous polyurethane dispersion were prepared by the reaction of hydroxyl-terminated poly(ethylene adipate), dimethylol propionic acid, 4,4'-diphenylmethane diisocyanate, and ethylene glycol. Formation of the dispersion was achieved by phase inversion of an acetone solution of the polyurethane with water, utilizing carboxylate anion groups as the internal emulsifying sites. The amount of acetone added has a large effect on the particle diameter (0.08 μm to 8.61 μm) and particle size distribution of the polyurethane emulsion. The storage stability was evaluated in terms of particle size and particle size distribution. The aqueous emulsion obtained with no use of acetone was sufficiently stable in storage at least over six months.

Keywords: particle size; particle size distribution; phase inversion; polyurethanes; shelf life stability

Introduction

From the view points of inherently environmental friendly process, workers' health and hygiene, and legislative requirements, polluting organic solvents (e.g., volatile organic compounds (VOCs)) that evaporate during the formulations of coatings cause a wide variety of air quality problems. In order to reduce or eliminate organic solvents from the formulations, organic solvents should be partially or completely replaced with an environmentally benign solvent (e.g., water), in the coating formulations to achieve small or no VOC content.[1] Polyurethane is widely used in the coating industry due to their excellent performance. The organic solvent-based (e.g., N,N'-dimethylformamide (DMF) and N,N'-dimethylacetamide (DMAc)) polyurethane (PU) will be

    DOI: 10.1002/masy.200451222

restricted in their traditional applications sooner or later because of the demands of environmental regulations. In contrast, aqueous PU dispersion would be one of candidates promising to replace them. An aqueous PU dispersion is a binary colloidal system in which PU particles are dispersed in an aqueous medium.

Aqueous-dispersible PU was introduced in open literature in late 1960s.[2] In order to produce high-molecular-weight aqueous PU dispersions, it is necessary to reduce the viscosity of mixing system. Various processes have been developed for the preparation of aqueous PU dispersions.[3–5] Among these processes, the acetone process represents one of the most popular processes,[6] in that acetone is inserted with respect to the PU formation, freely miscible with water, and a low boiling point solvent can be removed with ease. The addition of water to an acetone solution of PU ionomer followed by removal of the acetone results in aqueous PU dispersions.

A typical PU is produed by polyaddition of diisocyanates and diols so that urethane linkage occur in the polymer chain. An approach to making an aqueous-dispersible PUs is to incoporate hydrophilic segments into PU backbones by replacing a small portion of the aforementioned diols with component that contains aqueous-solubilizing functional groups such as carboxylate or sulfonate groups in its molecular structure. These components are often called internal emulsifiers or hydrophilic monomers.[7–8] Many research groups[9–13] have used the carboxylic acid containing low molecular weight diol dimethylol propionic acid (DMPA) as an internal emulsifier. The degree of hydrophilicity is one of the important factors determining the particle size of the PU particles. [10–13] In general, smaller particles result with increasing in hydrophilicity. In practice, PU emulsion is stored for a while prior use, therefore, storage stability is an important characteristic of an emulsion. In dilute emulsion systems, creaming velocity (v_{Stokes}) was described in Stokes terminal velocity equation[14–15]

$$v_{Stokes} = \frac{d_p^2 |\rho - \rho_o| g}{18\mu_c} \qquad (1)$$

where d_p is the diameter of emulsion particle, ρ and ρ_o are the density of water and emulsion, μ_c is the viscosity of water and g is the gravity acceleration velocity. Obviously, in the dilute emulsion system, diameter of emulsion particle has a significant effect on the creaming velocity. That is the smaller particle, the slower creaming velocity. The creaming velocity of an emulsion as defined by Stokes terminal velocity equation (Eq.1) gives only the rate of creaming of a single droplet. However, in a polydisperse system consisting of n_i droplets of radius r_i, the mass creaming rate $(\bar{u})$, has been defined as[16]

$$\bar{u} = \sum_i \frac{8\pi}{27\mu_o V} g\, n_i r_i^5 \left| \rho - \rho_o \right| \qquad (2)$$

where V is the total volume of the disperse phase, μ_o is the viscosity of emulsion.

Consequently, the shelf life colloidal stability of aqueous dispersion PUs is influenced by their particle size distribution.

Here, we describe the synthesis of PU having built-in carboxylic (COOH) groups as an internal emulsifier and the preparation of aqueous PU emulsions using a two-step process. Since, the shelf life stability is an important characteristic of an emulsion, the study focuses on the colloidal stability properties of aqueous dispersion PU. The effect of acetone content on the shelf life stability of the resulting emulsions was also investigated in terms of particle size, particle size distribution (PSD).

Experimental

Materials

Hydroxyl-terminated poly(ethylene adipate) (EGAA, $M_n = 1000$, Asahi Denka Kogyou) was dried by azeotropic distillation with toluene. The distillation was continued until no further distillate was collected. The moisture content of EGAA was determined by Karl Fisher titration (< 0.03 wt%). Ethylene glycol (EG, Nacalai Tesque), and triethylamine (TEA, Wako Pure Chem. Ind.) was dried according to standard methods just before use. N-Methyl-2-pyrrolidinone (NMP, Wako Pure Chem. Ind.) and acetone (Wako Pure Chem. Ind.) were dried over 4 Å molecular sieves. Dimethylol propionic acid (DMPA, Tokyo Kasei Kogyo) and 4,4'-diphenylmethane diisocyanate (MDI, Nippon Polyurethane Ind.) were used as received.

Preparation of Aqueous PU Dispersions

A series of aqueous PU dispersions were prepared with varying the amount of acetone. Their formulations are presented in Table 1. DMPA (ca. 3 wt% based on total solid PU) was dissolved in a minimum amount of NMP in a round bottom flask equipped with a stirrer, a thermometer, and an inlet of dry nitrogen. Molten EGAA and solid MDI (1.5 equiv) were added to the flask on a heating oil bath. Pre-polyaddition was carried out with mechanical stirring at 40 °C under a nitrogen atmosphere until the consumption of isocyanate (NCO) groups reached a theoretical amount estimated by assuming an equimolar reaction of the NCO and OH groups. This temperature and atmosphere was maintained to avoid any competitive side reactions, such as the formation of allophanates. The unreacted NCO groups were determined by a back-titration using di-*n*-butylamine. After the resulting viscous prepolymer was diluted with a given amount of acetone, EG (an equiv amount to the unreacted NCO groups) was added. The chain-extension reaction was continued at 50 °C until the IR spectra showed no stretching band due to the NCO groups at 2270 cm^{-1}. The COOH groups in the resulting chain-extended PUs were neutralized with TEA (1.2 equiv) in acetone at 50 °C for 1 h, and then water was gradually added with vigorous stirring (750 rpm) at 50 °C. The dispersion was stirred for further 30 min. Removal of acetone by a rotary vacuum evaporator gave a milky white aqueous PU emulsion or suspension, which contained about 30 wt% PU.

Table 1. Formulation for PU Dispersions.

Formulation	PUD1	PUD2	PUD3	PUD4
[NCO]/[OH]$_{EGAA}$/ [OH]$_{DMPA}$/[OH]$_{EG}$	1/0.49/ 0.17/0.34	1/0.49/ 0.17/0.34	1/0.47/ 0.20/0.33	1/0.48/ 0.18/0.34
DMPA (w/w% solid PU)	3	3	3	3
Neutralization (%)	120	120	120	120
Soft Segment (%)	60.88	61.11	59.06	60.21
Acetone (w/w% dispersion)	0	12	24	30

Characterizations

Infrared spectra were recorded on a JASCO model FT/IR-3 or A-202 infrared spectrometer. The particle size and particle size distribution (PSD) of PU emulsions were measured using of a laser diffraction particle size analyzer (Shimadzu Model SALD 2001) at ambient temperature. Because the PU emulsions exhibited in a skewed distribution function, the lognormal distribution was used for describing the particle size of PU emulsion. On the basis of the lognormal distribution, the geometric mean diameter (d_g) and geometric standard deviation (σ_g) were calculated according to eqs. (3) and (4), respectively.[17]

$$d_g = \exp\left(\frac{\Sigma(n_i \ln d_i)}{N}\right) \qquad (3)$$

$$\sigma_g = \exp\left(\frac{\Sigma n_i (\ln d_i - \ln d_g)^2}{N-1}\right)^{1/2} \qquad (4)$$

where n_i is the number of particles in group i, having a midpoint of size d_i, and where $N = \Sigma n_i$, that is, the total number of particles.

Results and Discussion

Preparation of Aqueous PU Anionomer Dispersions

Aqueous PU dispersions were prepared by two-step process, prepolymerization and chain extension, followed by subsequent neutralization of the carboxyl groups of DMPA with base. PU prepolymer with terminal isocyanate groups was prepared by the polyaddition of hydroxyl-terminated EGAA with an excess amount of MDI incorporating 3 wt% of DMPA. It is accepted that a minimum ionic content is required for the formation of stable aqueous PU dispersions, the actual amount depends on the type of ionic species and countercations (neutralizing agents) employed. Chen and Chen[9] found that fully neutralized PU anionomer with a minimum of about 0.8 wt% of COOH groups is required to form stable dispersions for at least 1 month. It has been reported that the average weight percentage of hydrophilic ionic site, rather than the average number of DMPA per polymer, is the critical factor for good dispersion.[18] The COOH content is defined as follows:

$$\text{COOH content (wt\%)} = \frac{\text{weight of COOH in solid PU}}{\text{total weight of solid PU}} \times 100 \qquad (5)$$

Because of the high melting point of DMPA (181–185 °C), minimum amount of water miscible cosolvent NMP was introduced to carry out a homogeneous reaction at 40 °C. NMP is often the preferred cosolvent because it acts as a process aid in film formation.[8] To test the stability of MDI in cosolvent NMP, blank runs were carried out in the reaction temperature. The results obtained for three hours reaction time are shown in Table 2.

No CO_2 gas bubbles were observed in prepolymerization of MDI, EGAA and DMPA, suggesting that no significant amount of water reacts with NCO groups and no reaction takes place between NCO groups and COOH groups in DMPA. The main advantage of DMPA is that the carboxylic acid is sterically hindered and so preferentially reacts into the backbone through the hydroxyl groups. It has been reported that in reaction of one equivalent of isocyanate with a hydroxy aliphatic acid, having carboxylic group, gives preferential reaction with the hydroxyl group.[19] Moreover, in prepolymerization stage, esterification would not take place because a higher temperature is required.[12,20] Therefore, the NCO groups of MDI react preferentially with the hydroxyl groups of EGAA and DMPA.

Table 2. Stability test of MDI in dried NMP at 40 °C.

Time (h)	[NCO] (mmol/g)	NCO unreacted (%)
0	5.157	100.0
0.5	5.157	100.0
3.0	5.136	99.6

The PU prepolymer was diluted with a given amount of acetone, and then chain-extended by the coupling between the unreacted NCO groups and EG. The post-polymerization PU thus obtained was composed of ternary diOH segments, such as EGAA, EG, and DMPA units. The DMPA unit has a potentially ionizable COOH group, which can act as an internal emulsifying agent. Since the average particle sizes decrease asymtotically as the TEA/DMPA mole ratio increase, because the carboxyl groups can become more dissociate. In addition, a TEA/DMPA mole ratio greater than 1.0 (more than 100% neutralization) is not affect the particle size.[11] Consequently, in order to secure a fully neutralization, the 1.2 times stoichiometric amount TEA was added to neutralize the built-in COOH groups of DMPA. As the ionomers formed, an increase of viscosity was attributed to the formation of microionic latices.[13] Lorenz et al.[21] have prove that in PU ionomer dispersions, the ionic sites COO^- is located substantially on the particle surfaces, while the hydrophobic chain segments form the interior of the particles. Thus, the aqueous PU emulsion was prepared by catastrophic phase inversion from the acetone solution to an aqueous emulsion by addition of water. The hydrophilic groups facilitated mutual repulsion between the particles provide particle stability. From the experiments, it was observed that the initial addition of water produced local turbidity that quickly disappeared with vigorous stirring. More addition of water made the bulk turbid, which was an indication of the formation of a dispersed phase. As water was continually added to a certain content, the phase inversion process was achieved. Particle stabilization from this structure can be described by the well known model of the diffuse electrical double layer.

Effect of Acetone on Particle Size

Addition of acetone to the prepolymer was suitable for reducing the viscosity and, in special, facilitating to disperse PU in water. The different aqueous PU emulsions were prepared with varying amounts of acetone from 0 to 30 wt% based on the total weight of the PU emulsion (30 wt% of PU in 70 wt% of water and NMP after removal of acetone). Figure 1 shows the particle size of the emulsions left for one day after the removal of acetone. Similar effect of the acetone content on the particle size was also visually observed as a transition from a bluish-white emulsion at 0 wt% of acetone to a milky white dispersion at high content of acetone. Apparently, the particle size increased with increasing the acetone content. A similar phenomenon has also been reported by other researchers.[6, 22–23] One possible explanation for this phenomenon lies in the retained acetone among the chains, even after the process of evaporation. Such acetone is probably adsorbed by the hydrophobic part of PU chain resulting in swelling of particle.

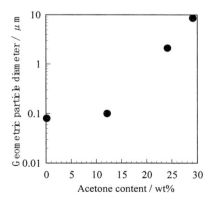

Figure 1. Effect of acetone content on the geometric particle diameter after 1-day storage.

The effect of acetone can be explained in terms of solubility parameter. According to the Hildebrand theory, the solvate power of a polymer-solvent medium can be estimated from $(\delta_1-\delta_2)^2$, where δ_1 and δ_2 are the solubility parameters for the solvent and polymer, respectively. The solubility of a polymer in a solvent is favored when $(\delta_1-\delta_2)^2$ is minimized, that is, when the solubility parameters of the two components are most closely matched. In this case the lower the difference between the solubility parameters of the hydrophobic part and the acetone, the higher the affinity between these two components. Literature data indicate that the solubility parameter for acetone[24] is 20.3 MPa$^{1/2}$ and that for polyurethane[25–26] ranges from 19.4 to 21.5 MPa$^{1/2}$. Thus, the affinity between the hydrophobic chain and acetone is high, so that the polyurethane chain configuration was easily to spread out in the presence of acetone in reaction process. Scaling arguments[27] can be used to interpret the dependence of the polymer size on the concentration of acetone. If $c \gg c^*$ the chain is ideal, and $R_g \propto c^{-1/8}$ (where c^* is the overlap concentration, c is the number of segment per unit volume and R_g is the radius of gyration of the polymer), therefore, a higher amount of acetone causes more swelling of the PU polymer.

Shelf Life of Aqueous PU Dispersions

Even though the pendant COOH-built-in PUs was capable of forming aqueous emulsions just after the preparation, some emulsions aggregated during storage at room temperature, resulting in a cream or precipitates. These phenomena seemed to associate closely with the amount of acetone used for preparing the emulsion. Recently, Keyvani[28] reported that distribution data, rather than average particle size numbers, can give a better indication of shelf life stability. Therefore, the storage stability of the PU emulsions prepared in 0 to 30 wt% of the acetone content was evaluated in terms of particle size distribution (PSD) change during the storage. In the case of a labile emulsion, the PSD should broaden and shift toward a larger particle size due to aggregation of the particles.[29] After one-day storage, the geometric particle diameter of the emulsion particle prepared with no use of acetone was 0.08 μm, and the size rarely changed even over six-month storage. Moreover, the emulsion showed almost no change of unimodal PSD during the storage, as shown in Figure 2(a). This indicates the high storage stability of this emulsion. However, the system was rather difficult to mix and disperse with water because of extremely high viscosity. To facilitate the mixing and dispersing process, acetone was added to the reaction system during the chain extension process. The particle size of the genetic emulsion particle prepared in 12 wt% of the acetone content was increased from 0.10 to 0.54 μm by about 4.4 times after one-month storage, resulting in a cream. Figure 2(b) shows the PSDs of emulsion prepared in 12 wt% of acetone, indicating a remarkable shift to a larger size during the storage. This phenomenon should be studied in further.

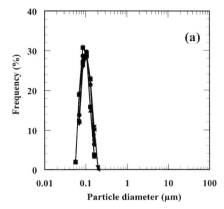 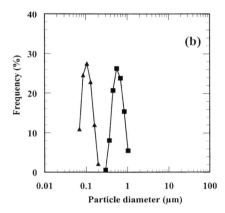

Figure 2. Time-dependence of the PSD for the aqueous PU emulsions; ▲ = after 1 day, ■ = after 1 month, ● = after 2 months, ◆ = after 3 months, ▼ = after 4 month, ◀ = after 5 months, ▶ = after 6 months; (a) 0 wt% of acetone, (b) 12 wt% of acetone.

On the other hand, the particle size of the emulsion particle prepared in 24 wt% of acetone was 2.11 μm after one-day storage. The appearances of second and third peaks of PSD were seen in storage for one month and later as shown in Figure 3. Similarly, the particle size of emulsion obtained in 30 wt% of acetone was 8.48 μm in unimodal PSD. However, second and third peaks of PSD were observed after one-month storage. Moreover, after two-month storage, the fourth peak was observed resulting in low stability.

238

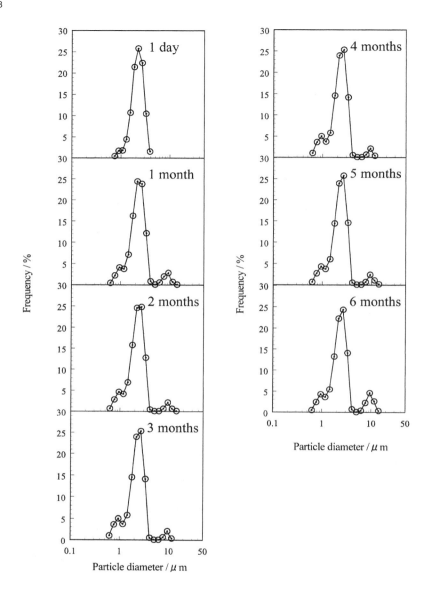

Figure 3. Time-dependence of the PSD for the aqueous PU emulsions; 24 wt% of acetone.

blowing agent concentration, with variation of nucleating agent concentration and with variation of catalyst concentration. Foams were characterized for cell window area, strut width and hydraulic resistance besides measurement of surface tension and bubble size in the polyol mixture.

Experimental

Raw Materials

The raw materials required to make water blown rigid polyurethane foam are polyol, isocyanate, catalysts, and surfactant. The polyol used was a sucrose based polyether polyol (DC 9911, Huntsman International, India). The polyol had a hydroxyl number 440 mg of KOH per gram of the polyol and an equivalent weight of 128 g/mole. The isocyanate was a polymeric diphenyl methane diisocyanate (MDI) (SUPRASEC 5005, Huntsman International). The isocyanate had an equivalent weight of 132 g/mole. The catalysts used were dibutyltin dilaurate (DBTDL) (Lancaster, England) and triethanolamine (TEA) (Spectrochem, India). Silica was procured from spectrochem, India. Tegostab B8404, silicone surfactant, was donated by Goldschmidt AG Germany. Atsurf 3315 and Cresmer B246M, polyether surfactants, were donated by ICI India. All materials were used as received without further purification.

Foam Formation

Polyol mixtures were prepared by stirring the 250 g of polyol containing different weight percentage (of the polyol) of water, catalysts and surfactant for half an hour. Then each polyol mixture was mixed thoroughly with predetermined amount of isocyanate for 15 seconds using a high-speed stirrer. The same speed was maintained for all formulations. Immediately the mixture was poured into a mould, which was kept at room temperature for 15 minutes. The foam was removed from the mould and all characterizations were done after curing the foams at room temperature for at least one day. All foams were made with an isocyanate index of 105. The densities of the foams were in the range of 140 to 165 kg/m^3.

Characterization

Surface tension of the polyol and solution of the surfactants (3 weight percent) in polyol were measured by Du Nouy Ring Tensiometer (Fisher Surface Tensiomat; Model 21). The average bubble size in the polyol containing one percent water and three percent surfactant was measured by using an optical microscope (Model BX60 Olympus). A drop of the polyol mixture after stirring for 20 minutes was taken over a glass slide and covered with a cover slip. Images were captured by an on-line CCD video camera (Model XC 77CE Sony) and frame grabber (Model Occulus MX, Coreco). All solutions were stirred at the same speed at which all foams were made. The average bubble diameter of the 400 bubbles was measured using an image analysis software (Image Pro Plus version 4.1). The cell window area and the strut width distributions were measured using the same transmission optical microscope. Thin slices, less than 1 mm, were cut from each foam sample and images were captured. The captured images were analyzed to measure cell window area and strut width by manually selecting the window area and strut width using the same image analysis software. Only those cell windows for which the whole window came into focus was taken. 70 measurements were made for each foam.

For hydraulic resistance measurement four samples of cubical shape with side 45 to 48 mm were cut from each foam. The weight of the samples was measured and then the samples were immersed in a sealed container filled with water. The container was connected to a pressure gauge. The pressure inside the container was raised to a specific value using a hydraulic hand pump. After one hour the foam samples were taken out and water from the surface of the samples was removed by a piece of cloth and again weighed. During this one hour the pressure was found to decrease due to water absorption by the foam samples. Thus at 5,10,15,20,25,30,40,50,55 minutes the pressure was checked and if there was any drop, the pressure was raised to the specified value. The percentage buoyancy loss was calculated by using the following formula,

$$percentage\ buoyancy\ loss = \frac{\rho'_f - \rho_f}{\rho_w - \rho_f} \times 100$$

where ρ_f, ρ'_f, ρ_w are the initial foam density, density of the foam after water absorption and the density of water respectively. The percentage buoyancy loss was measured for different hydraulic pressures in the range of 0 to 3 MPa.

Gel time was measured as follows. Initially the polyol mixture was prepared and then that was mixed with isocyanate for 15 seconds. A glass rod was vibrated into the reacting mixture. The point at which the mixture became stiff was considered as gel point. The gel time is the time between the start of mixing and the gel point.

Results and Discussion

a) Effect of Different Surfactant

Four foams were made for this part of the work. One of them was without any surfactant and the remaining three were with three different commercially available surfactants. In the formulations, the polyol contained 0.5% DBTDL, 0.5% TEA, 1% water and 3% respective surfactant. The surface tension of the polyol was 36.2 mN/m and with Atsurf 3315 it was 37.8 mN/m. The surface tension of the polyol with Cresmer B246M was 30 mN/m. Tegostab B8404 further lowered this value to 22 mN/m. The bubble size decreased with decrease in surface tension of the polyol. The average bubble diameter in the polyol was 57 ± 24 μm, in Atsurf 3315 54 ± 21 μm, in Cresmer B246M 44 ± 17 μm and in tegostab B8404 27 ± 9 μm. The distribution of cell window area of different foams has been shown in figure 1. Note the scales along x-axis for different foams are different. The average and standard deviation are given in each graph. The cell window areas of the foams with no surfactant and with Atsurf 3315 are very large and they are very widely distributed. Compare to these foams, the foams with Cresmer B246M and Tegostab B8404 have very small cell windows and their cell windows are very narrowly distributed. The cell windows of the foam with Cresmer B246M are bigger and the distribution is wider than that in the foams with Tegostab B8404. The strut width distribution pattern is similar to the distribution of cell window area. The average strut width in the foams with no surfactant and with Atsurf 3315 are 52 ± 9 and 52 ± 7 μm respectively. Whereas in the foams with Cresmer B246M and with Tegostab B8404 the average strut width are 41 ± 6 and 25 ± 4 μm respectively.

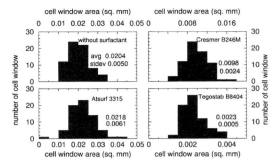

Figure 1. Effect of different surfactants on the distribution of cell window area. The average cell window area (avg) and the standard deviation (stdev) are indicated in each graph.

The above results indicate that a lower surface tension of the polyol mixture generates finer cells with a narrower distribution. The effect seems to be primarily due to a larger number of fine bubbles entrained during mixing. The effect of cell structure on hydraulic resistance of the foams is shown in figure 2. The error bars represent standard deviation for four samples. Lower

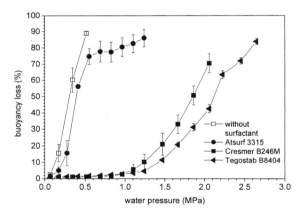

Figure 2. Effect of different surfactants on the hydraulic resistance of the foams. The error bar represents standard deviation for four samples.

the buoyancy loss higher is the foam's hydraulic resistance. The figure shows that foams with no surfactant and with Atsurf 3315 have very poor hydraulic resistance. The hydraulic resistance of the foams with Tegostab surfactant is more than that of the foam with Cresmer B246M. Thus hydraulic resistance of the foams increases with decrease in cell window area.

b) Effect of Surfactant Concentration at Different Water Concentration

For this part of the work nine foams were made. In the formulations, the polyol contained 0.5% DBTDL, 0.5% TEA. The amount of water and surfactant (Tegostab B804) are mentioned in the foam nomenclature. For example, foam XwYs was made with X% of water and Y% of surfactant.

The distribution of the cell window area is shown in figure 3. The average and standard deviation are given in each graph. At a particular water concentration, with increase in the surfactant concentration, the cell window area distributions become narrower and the average values become smaller. At 3% water, when surfactant concentration is increased from 5% to 7% (from foam 3w5s 3w7s) the cell windows become slightly bigger. At 3% and 5% surfactant concentrations, with increase in the concentration of water there is no appreciable change in the average values of cell window area. However the cell window area distributions become wider. At 1% surfactant concentration, when the water concentration is increased from 1% to 2% (from foam 1w1s to 2w1s), the average value of the cell window area increases. The distribution of cell window area also becomes wider. The cell windows become smaller and the distribution becomes narrower for further increase in water concentration from 2% to 3% (from foam 2w1s to 3w1s). The strut width distribution pattern is similar to that of the window area distribution pattern. The average strut width of the foams are in the range of 17 ± 3 to 27 ± 5 μm.

The effect of cell structure on hydraulic resistance of the foams is shown in figure 4. The error bars represent standard deviation for four samples. For constant surfactant concentration, the buoyancy loss increases with increase in the concentration of water, whereas for constant water concentration, the buoyancy loss decreases with increase in the concentration of surfactant. The efficiency of the surfactant concentration to lower the buoyancy loss increases with increase in the concentration of water. The penetration of water into the foam at high pressure essentially

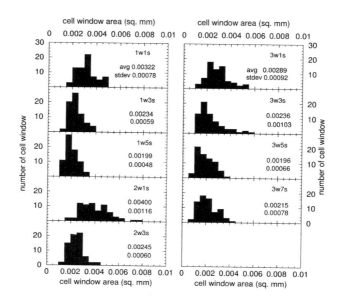

Figure 3. Effect of surfactant concentration at different water concentration on the distribution of cell window area. The average cell window area (avg) and the standard deviation (stdev) are given in each graph.

results from a sequential rupture of cell windows. The resistance to rupture of a cell window is higher for smaller and thicker windows. When the surfactant concentration is increased keeping water concentration constant, the windows are smaller and the hydraulic resistance of the foams increases. Higher amount of surfactant might have prevented the gas bubbles from coalescence at the initial stage of foam formation. When the water concentration is increased keeping surfactant concentration constant, the cell window area remains almost same however the hydraulic resistance decreases. This indicates that the cell window thickness decreases with increase in water concentration. This parameter was not measured in our experiments but a possible mechanism to explain why this should happen is as follows. When the water concentration was increased greater amount of carbon dioxide was generated, as the isocyanate index was kept constant. So the pressure inside the bubbles was higher. This higher pressure caused further drainage of the liquid. For this reason with increase in water concentration the windows become thinner and require less force to rupture.

c) Effect of Nucleating Agent Concentration

Micron sized silica particles were used as nucleating agent. The average particle diameter was 0.69 ± 0.16 μm. In the formulations the polyol contained 0.5% DBTDL, 0.5% TEA, 1% water. The amount of silica and surfactant (Tegostab B8404) are mentioned in the foam nomenclature. Foam ZsiYs was made with Z% of silica and Y% of surfactant.

The effect of nucleating agent on the distribution of cell window area and strut width is shown in figure 5. The average and standard deviation are mentioned in each graph. Note that 0si3s and 1w3s are the same foam as are 0si5s and 1w5s. The cell windows become smaller due

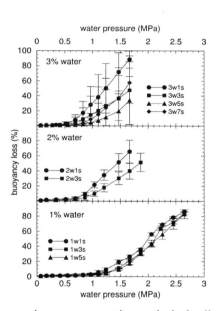

Figure 4. Effect of surfactant and water concentration on the hydraulic resistance of the foams. The error bar represents standard deviation for four samples.

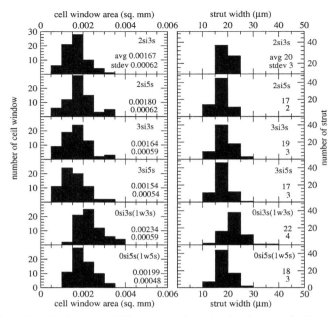

Figure 5. Effect of nucleating agent (silica particles) on the distribution of cell window area and strut width. The average values (avg) of cell window area and strut width and standard deviation (stdev) are given in each graph.

to addition of nucleating agent. However, there is no noticeable change either in the distribution or in the average values due to the variation of the concentration of nucleating agent and surfactant. The effect of silica particles as nucleating agent on buoyancy loss is shown in figure 6. The error bars represent standard deviation for four samples. Note that 0si3s and 1w3s are the same foam as are 0si5s and 1w5s. At higher pressures (0.80 MPa onwards) the buoyancy losses of the foams made with silica, are more than that of the foams made without silica. Thus, when silica particles are used as nucleating agent the cell windows are smaller but the buoyancy losses are more than that of the foams made without silica. This indicates cell windows in the foams with silica are much more thinner than the windows in the foams without silica. Independent measurements are not available to confirm this, however.

d) Effect of Variation of the Concentration the Catalysts

Foams for this part of the work were made with variation of the concentration of the catalysts, keeping total concentration (0.004 mole/100 gram of polyol) of the catalysts constant. In the formulations the polyol contained 1% water, 3% surfactant (Tegostab B8404). The proportion of DBTDL is mentioned in the foam nomenclature. For example, the proportion of DBTDL in the foam P_X is X.

Figure 7 shows the distributions of cell window area and the strut width. Note that the scales along x-axis in both window area and strut width distributions for the foam P_00 are different from rest of the distributions. The average and standard deviation of the measured quantity are mentioned in each graph. The cells in the foam P_00 are very coarse even observed by naked eye. The cell window area and strut width of this foam are most widely distributed and the average values of the cell window area and the strut width are also the largest. The distributions are much narrower and the average values are also smaller for the other foams. Increasing the mole fraction of DBTDL from 0.25 to 0.75 (from foam P_25 to P_75), both the cell window area and strut width distributions become narrower. However there are no appreciable differences both in the distributions and in the average values between foam P_75 and P_100.

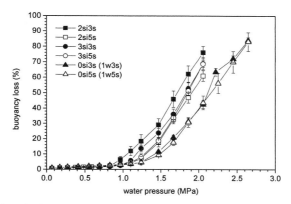

Figure 6. Effect of nucleating agent (silica particles) on the hydraulic resistance of the foams. The error bar represents standard deviation for four samples.

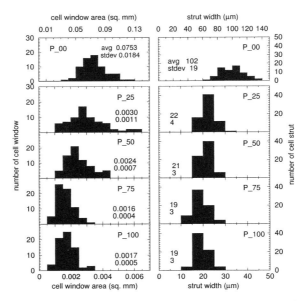

Figure 7. Effect of variation of catalysts concentration on the distribution of cell window area. The average cell window area (avg) and the standard deviation (stdev) are indicated in each graph.

The buoyancy losses of the foams at different water pressures are shown in figure 8. The error bars represent the standard deviation for four samples. At any particular water pressure, the buoyancy loss gradually decreases with increase in DBTDL proportion in the total concentration of the catalysts. The gel time was at a maximum for the foam P_00 (362 seconds) and minimum for the foam P_100 (29 seconds). The gel times of the other foams were as follows: 45 seconds for P_25, 37 seconds for P_50 and 32 seconds for P_75. The catalytic efficiencies of the different catalysts are different. For the foam P_00, the rate of network formation is much slower than any other foam, so the viscosity build up is not fast enough to withstand the pressure inside the bubbles, resulting in coalesence of the bubbles. For this reason the cell windows of the P_00 are very large. Due to low viscosity there is more cell window drainage and hence very thin cell windows. As a consequence the buoyancy losses of the P_00 are very large even at low pressures.

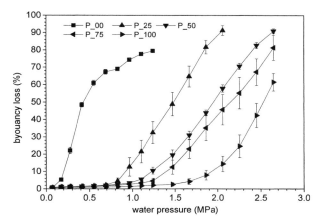

Figure 8. Effect of the variation of the concentration of the catalysts on the hydraulic resistance of the foams. The error bars represent the standard deviation for four samples.

As the proportion of DBTDL is increased the rate of network formation is accelerated and so there are lesser coalescence of the bubbles. In the final foams the cell windows are smaller. Figure 3 shows that the cell window areas and strut widths of the foams P_75 and P_100 are almost same. But the hydraulic resistance of the P_100 is more than that of P_75. This indicates the windows of P_100 are stronger, and hence thicker than the windows of P_75. The gel time of P_100 is less than P_75, so the viscosity built up is faster for P_100 and thus there is less time for cell windows drainage. This would lead to thicker windows for P_100.

Conclusion

Cell size in rigid polyurethane foam can be regulated according to the requirement. Proper selection of the surfactant that can lower the surface tension of the polyol and thereby facilitate generation of large number of small bubbles, giving foam with very small cell size. Cell sizes can be reduced with a narrow distribution by increasing the concentration of the surfactant. Using nucleating agent cell size can be reduced but the windows become thinner. Finally, by adjusting the rate of the blowing and network formation, foam can be made with narrowly distributed cells having thicker cell windows.

254

[1]   G. Oertel, *"Polyurethane Handbook"*, Hanser Publisher, Munich 1985, Chap. 1, p. 6.

[2]   M. Szycher, *"Handbook of Polyurethanes"*, Boca Raton, CRC Press LLC, New York 1999, p. 1.

[3]   G. Oertel, *"Polyurethane Handbook"*, Hanser Publisher, Munich 1985, Chap. 1, p. 99.

[4]   M. Szycher, *"Handbook of Polyurethanes"*, Boca Raton, CRC Press LLC, New York 1999, p. 8.

[5]   X. D. Zhang, C. W. Macosco, H. T. Davis, A. D. Nikolov, D. T. Wasan, *J. Colloid Interface Sci.* **1999**, *215*, 270.

[6]   H. J. Kollmeier, H. Schator, *J. Cell Plast.* **1985**, July-August, 239.

[7]   M. Szycher, *"Handbook of Polyurethanes"*, Boca Raton, CRC Press LLC, New York 1999, p. 8.

[8]   S. G. Luo, H. M. Tan, J. G. Zhang, Y. J. Wu, F. K. Pei, X. H. Meng, *J. Appl. Polym. Sci.* **1997**, *65*, 1217.

[9]   N. C Hilyard, A. Cunninggham, *"Low Density Cellular Plastics: Physical Basis of behaviour"*, Chapman & Hall, 1994, Chap. 2. p. 27.

[10]   K. C. Frisch, J. H. Saunders, *"Plastic Foams"*, Marcel Dekker INC, New York 1972, Part I, Chap. 2., p. 27.

[11]   A. Prociak, J. Pielichowski, T. Sterzynski, *Polymer Testing* **2000**, *19*, 705.

[12]   A. Swaminathan, D. V. Khakhar, *Cellular Polymers* **2000**, *19*, 103.

Macromol. Symp. **2004**, *216*, 255-263

Structure and Properties of Biodegradable Polymer-Based Blends

Zhaobin Qiu,[1] *So Fujinami,*[1] *Motonori Komura,*[2] *Ken Nakajima,*[1] *Takayuki Ikehara,*[3] *Toshio Nishi**[1]

[1] Department of Organic and Polymeric Materials, Graduate School of Science and Engineering, Tokyo Institute of Technology, 2-12-1 Ohokayama, Meguro-ku, Tokyo 152-8552, Japan
E-mail: tnishi@polymer.titech.ac.jp
[2] Department of Applied Physics, School of Engineering, The University of Tokyo, 7-3-1, Hongo, Bunkyo-ku, Tokyo 113-8656, Japan
[3] Department of Applied Chemistry, Faculty of Engineering, Kanagawa University, 3-27-1, Rokkakubashi, Kanagawa-ku, Yokohama 221-8686, Japan

Summary: We have recently investigated the structure and properties of biodegradable polymer based blends. The biodegradable polymers studied in our work include poly(3-hydroxybutyrate) (PHB), poly(3-hydroxybutyrate-hydroxyvalerate) (PHBV), poly(butylene succinate) (PBSU), poly(ethylene succinate) (PES) and poly(ε-caprolactone) (PCL). The miscibility, morphology, crystallization and melting behaviour of biodegradable polymer based blends have been studied extensively, which were reviewed briefly in this manuscript.

Keywords: biodegradable; blends; crystallization; miscibility; morphology

Introduction

Biodegradable polymers have received considerable attention in the last two decades due to their potential applications in the fields related to human life such as environmental protection and ecology. According to the difference in the preparation methods, biodegradable polymers can be classified into two types. One is the biosynthetic polymers, such as bacterial polyhydroxyalkanoates (PHAs). Among them, PHB and PHBV are probably the most extensively studied biodegradable thermoplastic polymers. The other type is the chemosynthetic polymers such as aliphatic polyesters. PBSU, PES and PCL are three of them.

          DOI: 10.1002/masy.200451224

In order to improve the properties and extend the application fields of the biodegradable polymers, polymer blending is an effective and convenient way. [1-15] In this manuscript, we briefly reviewed our recent research on the miscibility, morphology, crystallization and melting behaviour of polymer blends based on the five biodegradable polymers mentioned above. [8-15]

Miscibility of polymer blends can be divided into three types: namely, completely miscible, partially miscible and completely immiscible. Furthermore, in terms of the crystallizability of the components binary polymer blends can be classified into three types: namely amorphous/amorphous, amorphous/crystalline and crystalline/crystalline polymer blends. Blends in which both components are crystalline polymers have received much less attention than fully amorphous or amorphous/crystalline systems. Only a small number of works have been reported on the miscible polymer blends of two crystalline polymers with different chemical structures till now. However, they may be of considerable technological interest and offer the possibility of investigating crystallization and morphological behavior related to blend miscibility. On the other hand, it is also possible to determine how the crystallinity of one component affects the morphology, crystallization process, and properties of the other in the case of binary immiscible blends of two crystalline polymers.

We have recently done some research on the misciblity and crystallization behaviour in biodegradable polymer based blends. They include miscible crystalline/crystalline polymer blends, immiscible crystalline/crystalline polymer blends and miscible crystalline/amorphous polymer blends.

Miscible Crystalline/Crystalline Polymer Blends

PES/poly(ethylene oxide) (PEO) and PBSU/PEO blends are two new ideal models of binary miscible blends of two crystalline polymers. [8, 10-11]

PES/PEO blends were preapred with mutual solvent chloroform. PES and PEO were miscible blends of two crystalline polymers, which were crystallized at 50 °C to study the possibility of simultaneous crystallization of the two components by optical microscopy (OM). It was found that down to PES/PEO = 40/60 PES crystallized first and filled the whole space before PEO could nucleate. However, for PES/PEO =20/80, PES and PEO crystallized simultaneously and the formation of interpenetrated spherulites of PES by PEO was observed. It was proposed that the density of lamellae in PES spherulites was lower than that in PEO spherulites, and PEO spherulites continued to penetrate into PES spherulites on contact. [8]

Interpenetrated spherulites are occasionally formed in few miscible pairs of two crystalline polymers. Spherulites of one component continue to grow in the spherulites of the other component after they contact with each other. In our previous work, we reported the interpenetrated spherulites formation process of poly (butylene succinate) (PBSU)/poly (vinylidene chloride-co-vinyl chloride) (PVDCVC) blends and poly(ester carbonate) (PEC)/poly(L-lactic acid) (PLLA).[3-7] From the previous studies, we propose that the important factors in realizing interpenetrated spherulites are the difference in the lamella population density in the different spherulites of the two components, the sufficient amount of the melt of one component inside the spherulites of the other component, and the simultaneous spherulitic growth of both components.

As shown in Figure 1, PES and PEO crystallized simultaneously with PEO spherulite generally growing faster than PES spherulite. Instead of growth being arrested, the PEO spherulite continued to crystallize inside the PES spherulite and the growth front of the PEO type spherulite became distorted when it reached the PES type spherulite. Meanwhile, the brightness increased in the part of PES spherulite where the crystallization of PEO had occurred, and did not change where the crystallization of PEO had not occurred. These results indicated that PES spherulite was penetrated by PEO spherulite. However, it can also be observed that the outline of the brighter area is not consistent with the shape of the rest of the PEO spherulite, and the brighter area seems too large. It can be reasoned that the PEO fibrils grew faster inside the PES spherulite because the PEO content in the amorphous regions of the PES spherulite would be expected to be higher than the nominal melt concentration due to rejection of PEO from PES crystals. On the other hand, the PES spherulites stopped advancing towards the PEO spherulites.

PBSU/PEO blends were prepared with mutual solvent chloroform. The misciblity and crystallization behaviour were studied by DSC and OM. Experimental results indicate that PBSU is miscible with PEO as shown by the existence of single composition dependent glass transition temperature over the entire composition range. In addition, the polymer-polymer interaction parameter, obtained from the melting depression of the high-T_m component PBSU using the Flory-Huggins equation, is composition dependent, and its value is always negative. This indicates that PBSU/PEO blends are thermodynamically miscible in the melt. The morphological study of the isothermal crystallization at 95 °C (where only PBSU crystallized) showed the similar crystallization behavior as in amorphous/crystalline blends. Much more attention has been paid to the crystallization and morphology of the low-T_m component PEO. The two components crystallized sequentially not simultaneously when the blends were quenched from the melt directly to 50°C (one-step crystallization), and the PEO spherulites crystallized within the matrix of the crystals of the preexisted PBSU phase. Figure 2 shows the bright PEO spherulites growing in the matrix of the PBSU spherulites, which formed during the quenching process from the crystal-free melt to the crystallization temperature 50 °C. Before the temperature reached 50 °C, PBSU spherulites appeared and filled the whole space. Bright PEO spherulites thus nucleated and grew in the matrix of the PBSU spherulites after the crystallization temperature 50 °C was reached, which was lower than the melting point of PEO. [10]

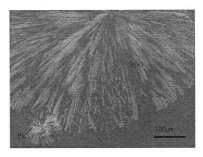

Figure 1. Formation of interpenetrated spherulites for PES/PEO=20/80 blend crystallized at 50 °C.

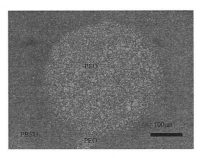

Figure 2. Bright PEO spherulites growing in the matrix of previously formed PBSU spherulites.

Acknowledgement

Z. Qiu thanks the Japan Society for the Promotion of Science for providing the fellowship and the grant-in-aid (P01278) to do this research at the University of Tokyo and Tokyo Institute of Technology.

[1]   C. S. Ha , W. J. Cho, *Prog Polym Sci* **2002**, *27*, 759.
[2]   J. C. Lee, H. Tazawa, T. Ikehara, T. Nishi, *Polymer J.* **1998**, *30*, 327.
[3]   J. C. Lee, H. Tazawa, T. Ikehara, T. Nishi, *Polymer J.* **1998**, *30*, 780.
[4]   T. Ikehara, T. Nishi, *Polymer J.* **2000**, *32*, 683.
[5]   Y. Terada, T. Ikehara, T. Nishi, *Polymer J.* **2000**, *32*, 900.
[6]   S. Hirano, Y. Nishikawa, Y. Terada, T. Ikehara, T. Nishi, *Polymer J.*, **2002**, *34*, 85
[7]   T. Ikehara, Y. Nishikawa, T. Nishi, *Polymer* **2003**, *44*: 6657.
[8]   Z. B. Qiu, T. Ikehara, T. Nishi, *Macromolecules* **2002**, *35*, 8251.
[9]   Z. B. Qiu, T. Ikehara, T. Nishi, *Polymer* **2003**, *44*, 2503.
[10]   Z. B. Qiu, T. Ikehara, T. Nishi, *Polymer* **2003**, *44*, 2799.
[11]   Z. B. Qiu, T. Ikehara, T. Nishi, *Polymer* **2003**, *44*, 3095.
[12]   Z. B. Qiu, T. Ikehara, T. Nishi, *Polymer* **2003**, *44*, 3101.
[13]   Z. B. Qiu, T. Ikehara, T. Nishi, *Polymer* **2003**, *44*, 7519.
[14]   Z. B. Qiu, T. Ikehara, T. Nishi, *Polymer* **2003**, *44*, 7749.
[15]   Z. B. Qiu, T. Ikehara, T. Nishi, *Polymer* **2003**, *44*, 8111.

Macromol. Symp. **2004**, *216*, 265-279   265

Effects of Compositions and Processing Variables on Barrier and Mechanical Properties of Liquid Crystalline Polymer/Polyethylene Blend Films

Tatiya Trongsatitkul,[1] *Duangdao Aht-Ong,*[2] *Wannee Chinsirikul**[1]

[1] National Metal and Materials Technology Center, National Science and Technology Development Agency, Thailand Science Park, Pathumthani, Thailand
E-mail: wanneec@mtec.or.th
[2] Department of Materials Science, Faculty of Science, Chulalongkorn University, Bangkok, Thailand

Summary: Cast films of liquid crystalline polymer (LCP) and low density polyethylene (LDPE) blends have been produced and investigated. Effects of LCP content and processing parameters, i.e., processing temperature profile, screw speed, and post-die drawing, on morphology and O_2 barrier property are presented. Increasing processing temperature and LCP content tend to enhance aspect ratios (L/D) of the LCP dispersed phase and at the same time influencing LCP structure. These effects are clearly observed when LCP content is increased from 10 % to 30 % by wt. At high temperature profiles, LCP morphologies are presented in a more or less 'ribbon' or 'tape' like structure together with a common LCP fibrillar structure. Films of 10% and 30% LCP produced at two optimum temperature profiles show a noticeable proportion of LCP tape-like structure and interestingly high barrier properties of ~1.6 and 5.5 times that of the neat LDPE films. High barrier characteristics of such LCP/PE blend films are indicated by low oxygen transmission rate values. Apart from processing temperature effect, increases of screw speed result in films having smaller aspect ratios for both LCP fibers and ribbons; films also exhibit poorer barrier and mechanical properties. However, post-die drawing clearly demonstrates a positive effect in improving aspect ratios of LCP domains and the resulting films' moduli. Effects of post-die drawing on enhancing films' barrier properties become more pronounced at high LCP content. By comparing with the neat LDPE film (30 μm thick) having modulus of ~180 MPa and OTR of ~11000 cc/m². day, the developed LCP/PE films containing 30 wt% LCP show remarkably high modulus values of ~1100 MPa with low OTR of ~2000 cc/m².day.

Keywords: barrier property; immiscible blend; liquid crystalline polymer; morphology control; oxygen transmission rate

          DOI: 10.1002/masy.200451225

Introduction

Development of polymer products with desirable barrier properties by blending technique has been of considerable interests in the last decade [1-3]. Multiphase monolayer films have been considered as alternatives to multilayer films. Multilayer films are produced using a rather complex process based on lamination and co-extrusion. Furthermore, difficulties in recycling multilayer products have also been major drives towards developing polymer blend films with recyclability.

Regarding barrier property improvement via blending, high barrier polymers such as EVOH, PEN, Nylon and LCP can be incorporated into another polymer matrix to create tortuous pathway for enhancing gas traveling charateristics. The key to achieve desirable barrier performance is morphology control [2, 3]. It has been suggested that laminar structure is preferable for high barrier requirement [1, 2, 4]. EVOH and Nylon [3] have been considered interesting candidates, but some limitations still exist. For EVOH, effective reduction of O_2 permeability can be obtained only in dry atmosphere. Therefore, this research proposes to study effectiveness of another candidate, a super-barrier LCP, in enhancing barrier properties of commonly used LDPE through polymer blending. Challenges involve trying to synergize unique self-reinforcing property and excellent barrier characteristics of the LCP with desirable flexibility of LDPE in order to produce practical films with reasonable cost.

Experimental

Materials: LCP used was Vectra type, namely LKX1107 with a low melting point of ~220 °C, kindly supplied by Ticona. The polymer matrix LDPE was an extrusion grade, LD1902F, obtained from Cementhai Chemicals Co., Ltd. Melt flow index of the chosen LDPE is 2 g/10min and its melting temperature is ~108 °C. The compatibilizer and antioxidant used were Nucrel 0903 and Irganox 1076; their contents were kept constant for all blends at 0.5 wt% and 0.1 wt%, respectively.

Blending and Pelletization

Various contents of dried LCP, LDPE, and additives were mechanically blended and extruded using Haake single screw extruder having a mixing configuration (19 mm screw diameter and L/D ~25). Blends were extruded through a rod die of 3 mm diameter and pelletized.

Temperature setting from feed zone to die zone was 170-200-230-225 °C. Screw speed of 75 rpm was kept constant for pelletization. Pellets were also dried prior to being used for film fabrication.

Cast Film Preparation

Cast films of LCP/LDPE blends were extruded through a slit die of 15 cm width and 0.8 mm die gap. Films of various LCP contents in a range of 5-40 % by wt were produced at different processing temperature profiles (T1-T6) with different screw speeds and degrees of post-die drawing (draw ratios), as summarized in Table 1. It should be noted that all films were fabricated using a constant screw speed of ~12 rpm. Influence of temperature profiles (T1-T6) on films' structures and barrier properties were investigated for blends containing 10 %LCP. Selected two temperature profiles of T2 and T6 providing noticeable improvements in films' barrier characteristics were then used for further studies on effects of LCP content, screw speed and post-die drawing.

Morphological Studies: Morphologies of LCP/LDPE blends were investigated using polarized light microscope and scanning electron microscope (SEM). In this study, blend samples were placed between two glass slides and heated to a temperature of ~160°C in order to melt LDPE while the LCP structure generated during processing could be revealed. Film specimens for SEM studies were fractured in liquid nitrogen. Fractured surfaces were then coated with gold prior to examination.

Tensile Testings: The tensile properties were determined on an Instron 4502 according to the ASTM D-882 for thin plastic sheet. Five specimens of each blend film in both MD and TD were tested and average value was reported.

O$_2$ Transmission Rate (OTR): OTR values of film samples were determined at 23 °C, 0 %RH using O$_2$ Permeation Analyzer (Illinois, Model 8500). At least 3 replicates of 100 mm^2 area were tested and averaged.

Table 1. Compositions and processing conditions for LCP/LDPE films extrusion.

Temperature Profile and Screw Speed Variation			
LCP/LDPE Composition	Temperature Profiles	Draw Ratio	Screw Speed
	Extruder Zone1-3 and Die	(Post-Die Drawing)	(rpm)
10/90	(T1) 180-200-225-220	1.5	12
	(T2) 180-200-230-225	1.5	12, 20, 40, 60
	(T3) 180-200-235-230	1.5	12
	(T4) 180-200-240-235	1.5	12
	(T5) 180-200-270-270	1.5	12
	(T6) 180-200-300-300	1.5	12, 20, 40, 60
LCP Content and Post-Die Drawing Variation			
LCP/LDPE Composition	Temperature Profiles	Draw Ratio	Screw Speed
	Extruder Zone1-3 and Die	(Post-Die Drawing)	(rpm)
5/95	(T2) 180-200-230-225	1.5	12
	(T6) 180-200-300-300	1.5	
10/90	(T2) 180-200-230-225	1, 1.5, 2, 3	12
	(T6) 180-200-300-300	1, 1.5, 2, 3	
15/85	(T2) 180-200-230-225	1.5	12
	(T6) 180-200-300-300	1.5	
20/80	(T2) 180-200-230-225	1.5	12
	(T6) 180-200-300-300	1.5	
30/70	(T2) 180-200-230-225	1, 1.5, 2, 3	12
	(T6) 180-200-300-300	1, 1.5, 2, 3	
40/60	(T2) 180-200-230-225	1.5	12
	(T6) 180-200-300-300	1.5	

Results and Discussion

Film Morphologies:

Effect of Processing Temperature Profiles

For temperature profile study, all blend samples had the same LCP content of ~10 wt%. Temperature profiles of the first and second zones were kept constant at 180 °C and 200 °C, respectively, while the third zone and die temperatures were varied from 220 °C to 300 °C, identified as profiles T1-T6 (**Table 1**). Initial temperature profile setting of T1 was utilized based on the previous work on LCP/PE blends [5], where good fibers with high aspect ratios of ~100 and dramatic improvement in films' moduli could be achieved. In the current work, further increases in temperature profiles from T2 to T6 were carried out to vary viscosity ratios of the blends and to study their effects on morphology, mechanical and barrier performances. From SEM micrographs of fractured samples, Figures 1 (a)-(f), it is apparent that LCP fibrillar structure can be formed in the films produced at all 6 temperature profiles. However, these films contain different amount of LCP fibers with various fiber sizes and aspect ratios. In general, obtained fibers have aspect ratios of higher than 100 giving rise to effective fiber reinforcements and improved mechanical properties. When processing temperatures are increased, LCP fibers tend to be larger in width and appear to be more 'tape-like' structure, as seen in Figures1 (e) and (f). Such changes in LCP morphologies upon increasing processing temperature can be attributed to changes in viscosity ratios of the blends' components [4]. Effect of temperature profiles on the formation of LCP tape-like structure can be more apparent at higher LCP content of 30 wt% as described in the next section.

Effect of Screw Speed

Screw speeds of extruder were varied in order to study influence of shear rate on the LCP dispersed phase. The study was carried out by producing films of 10%LCP/LDPE with different screw speeds of 12, 20, 40, and 60 rpm at the two selected temperature profiles of T2 and T6. Note that draw ratios were kept constant at 1.5. In order to retain a constant draw ratio while increasing screw speed, take off speed must also be adjusted accordingly. From Figures 2, it is generally seen that fibrillar structure of LCP in films produced at T2 as well as ribbon-like structure of LCP in films produced at T6 still mainly exist. Nevertheless, both fibers and ribbons in the blends tend to be finer and shorter at high screw speeds. These observations could be due to the combined effects of the high shear rate and high take-up speed during processing under high screw speed conditions. Therefore, the occurrence of LCP fiber or tape breakage can take place.

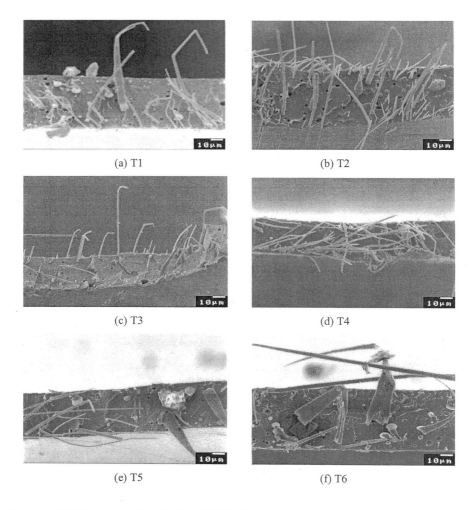

(a) T1

(b) T2

(c) T3

(d) T4

(e) T5

(f) T6

Figure 1. SEM micrographs of 10 wt%LCP blend films produced at difference temperature profiles T1-T6 (a-f) (750X magnification).

(a) T2, Screw speed 12 rpm   (b) T6, Screw speed 12 rpm

(c) T2, Screw speed 40 rpm   (d) T6, Screw speed 40 rpm

Figure 2. SEM micrographs showing effect of screw speed on morphology of 10 wt%LCP/LDPE blend films produced at T2 and T6 (750X magnification).

Effect of LCP Content

Films with various LCP contents ranging from 5-40 wt%LCP were produced using two temperature profiles, T2 and T6. For films having low LCP content of 5 wt% produced at T2, LCP phase is mainly present in droplet and ellipsoid. As LCP content is increased, there is a greater tendency in forming LCP fibrillar structure (Figure 3 (a) and (b)). However, when films are produced using higher temperature profile (T6), fibrillar and tape like structure of LCP are observed even at low LCP content of 5 wt% (Figure 4 (a) and (b)). These results are possibly attributed to different viscosity ratios of the blends and initial droplet size of LCP at the two processing conditions. Based on the Taylor's theory, there should be a required critical volume of LCP phase for effective fiber formation [3, 4]. Current findings in the formation of LCP tape-like structure at T6 may indicate a possible interplay of several key factors including viscosity ratio of the blends, initial LCP droplet size and its critical volume.

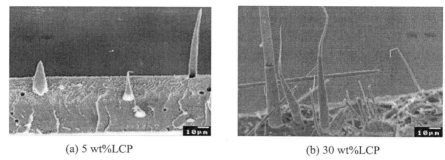

(a) 5 wt%LCP   (b) 30 wt%LCP

Figure 3. SEM micrographs of films produced at T2 (750X magnification).

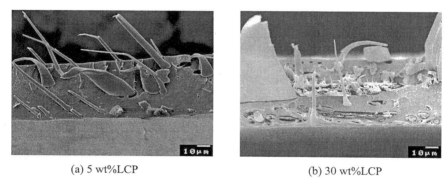

(a) 5 wt%LCP   (b) 30 wt%LCP

Figure 4. SEM micrographs of films produced at T6 (750X magnification).

Effect of Post-Die Drawing

Drawing within the fabrication process also influences the development of morphology in the blends. As earlier discussed, structure of LCP domain prior to deformation (e.g., droplet size and its volume) depends strongly upon both processing temperature and LCP content. While post-die drawing can enhance fibrillation and orientation, it is also found that extents of drawing directly affect final LCP fiber morphologies. Drawing PE blends film containing 10 wt%LCP, produced at T2, to a draw ratio (DR) of 1.5 results in numerous thin LCP fibrils. However, further drawing to a higher DR of 3 leads to fibril breakage (Figures 5 (a) and (b)). In contrast, for film with

30 wt%LCP produced at T2, LCP fibrils with high aspect ratios are still present at a high DR of 3 (Figures 5 (c) and (d)). In the case of films prepared at higher temperature profile of T6, SEM micrographs (Figure 6) reveal relatively large LCP fiber diameter of 5-10 μm with high aspect ratios and interesting LCP tape-like structure of ~10 - 30 μm width in all film compositions with various draw ratios.

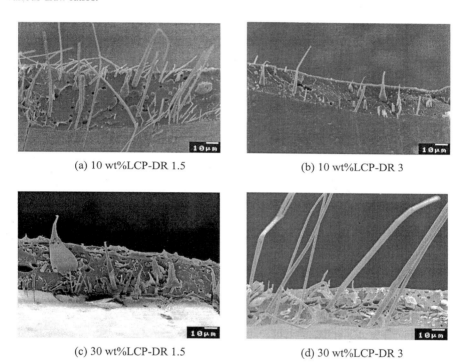

(a) 10 wt%LCP-DR 1.5   (b) 10 wt%LCP-DR 3

(c) 30 wt%LCP-DR 1.5   (d) 30 wt%LCP-DR 3

Figure 5. SEM micrographs showing drawing effects on films produced at T2 (750X magnification).

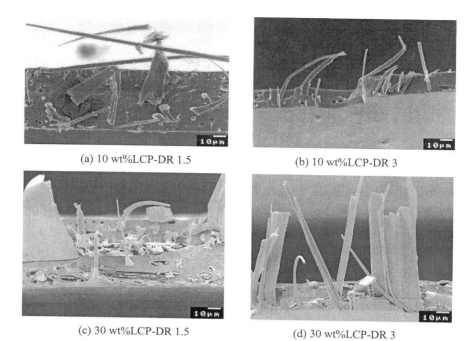

(a) 10 wt%LCP-DR 1.5

(b) 10 wt%LCP-DR 3

(c) 30 wt%LCP-DR 1.5

(d) 30 wt%LCP-DR 3

Figure 6. SEM micrographs showing drawing effects on films produced at T6 (750X magnification).

Film Properties:

Tensile Properties

LDPE films containing 10 wt%LCP show significant improvements in tensile properties; an increase of ~800 % in modulus and ~25 % in yield stress are observed as compared to those of LDPE (Figure 7). Varying temperature profiles from T2 to T6 results in slight increases in both properties, presumably due to better reinforcing effects of developed LCP structures at higher temperatures. Increase of screw speed tends to reduce film modulus which can be a result of LCP fiber/ribbon breakage (Figure 8). As briefly discussed, drawing during film extrusion can effectively enhance fibrillation and orientation of LCP, and correspondingly the film modulus [7]. As seen in Figure 9, this is essentially true for the LCP/LDPE films produced at the temperature profile of T6; thinner films with higher post-die draw ratios possess greater modulus values. The resulting thin films (~30 μm) having modulus of ~1100 MPa can be obtained. However, films produced at lower temperature profile of T2 exhibit an optimum thickness of ~40 μm or with a draw ratio of ~2, below which modulus decreases dramatically caused by fibrils breakage upon further stretching. The LCP fibril breakage is illustrated in Figures 5 (b).

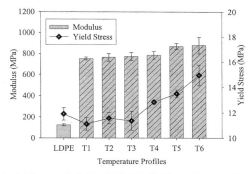

Figure 7. Modulus and yield stress in MD of 10 %LCP blend films produced at T1-T6 vs. LDPE film produced at T1.

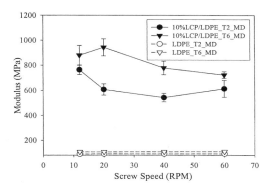

Figure 8. Modulus in MD of 10 %LCP blend films produced at T1-T6 vs. LDPE film as a function of screw speed.

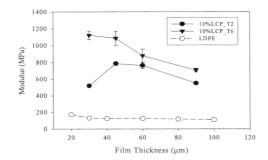

Figure 9. Modulus in MD of 10%LCP blend films produced at T2 and T6 vs. LDPE film produced at T1 (Effect of post-die drawing).

O₂ Transmission Rate (OTR)

OTR results of 10 %LCP/LDPE films processed at different temperatures are illustrated in Figure 10. There is no general trend of how O_2 permeability is affected by processing temperature. Nevertheless, noticeable decrease in OTR values of approximately 15 % and 40 % are observed in films produced at the conditions T2 and T6. Studying effects of screw speed on films' OTR reveal that the films containing 10%LCP posses an observable low OTR of ~3000 cc/m^2.day when produced under T6 condition with a low screw speed of 12 rpm (Figure 11). OTR value of this 10%LCP/LDPE film is significantly lower than that of the neat LDPE produced under a similar condition. Such low OTR property can be largely attributed to the developed LCP tape-like structure in the film as previously described. For 10%LCP/LDPE films produced at higher screw speeds of 20 - 60 rpm and at both T2 and T6, all films show high OTR values in a range of 5500 – 6 000 c c/m^2.day, which appear to be closed to the OTR of LDPE. Effects of post-die drawing with various film thicknesses together with effects of LCP content on films' OTR are presented in Figure 12. As expected, thinner films show higher OTR values. It is interesting to note that while OTR values of 10 %LCP produced at T2 show only a slight deviation from OTR of the LDPE matrix at all thicknesses, films containing the same amount of LCP (10 wt%), but produced at higher temperature of T6, exhibit noticeably lower OTR values. Furthermore, films containing high content of LCP (30 wt%) and produced at the temperature profile of T6 show even further decrease in OTR. In other words, the resulting films of 30 μm produced at T6 possess a remarkable 5 - 6 folds increase in oxygen barrier property over the LDPE matrix. OTR values of these films, containing 30 %LCP produced at T6, also become less thickness dependent. Significant enhancement in oxygen barrier characteristics of LDPE containing 30 %LCP, as indicated by low OTR value, can be attributed to the developed structure of LCP phase in a more or less 'tape-like' feature. It is possible that such LCP tapes, dispersed in the LDPE matrix, play a role in enhancing pathway or distance for gas traveling across the film thickness.

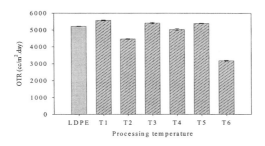

Figure 10. OTR of 10 %LCP blend film produced with different processing temperatures T1-T6 vs. LDPE film produced at T1 (all films have thicknesses of ~60 μm).

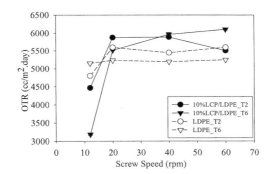

Figure 11. Effect of screw speed on oxygen transmission rate of 10%LCP/LDPE films produced at temperature profile T2 and T6 as compared to LDPE films.

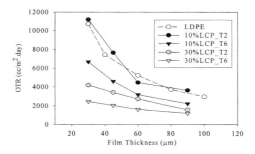

Figure 12. OTR of 10 and 30 %LCP blend films produced at T2 and T6 vs. OTR of LDPE films.

Conclusion

The effects of LCP content and processing parameters i.e., temperature profiles, screw speed, and post-die drawing on films' morphology, tensile and barrier properties have been reported. Films with effective LCP fibrillar structure (L/D >100) can be obtained at all six temperature profiles (T1-T6). However, increasing processing temperature can give rise to larger LCP fibers of high aspect ratios together with LCP 'tape-like' structure. Increasing screw speeds adversely result in films with shorter LCP fibers or tapes, poorer barrier and tensile properties due to LCP fiber or tape breakages. Post-die drawing clearly enhances fibrillation and orientation of LCP. Varying draw ratios (DR) of the 10%LCP/LDPE films produced at T6 from 1.5 to 3 can lead to a significant enhancement in fibers' aspect ratios and films' moduli of ~60 %. Barrier properties of LCP/LDPE films can be largely improved by developing LCP 'tape-like' structure under appropriate processing conditions or at the high temperature profile of T6. Thin films of 30 μm containing 10 %LCP and 30 %LCP show interestingly low OTR values o f ~4000 and ~2000 cc/m^2.day while obtaining reasonably high modulus values of ~600 and 1100 MPa as compared to the neat LDPE with OTR of ~11000 cc/m^2.day and a relatively low modulus of ~200 MPa.

Acknowledgements

The authors would like to acknowledge MTEC and staffs for technical supports, and Ticona for providing LCP used in this research. The first author is grateful to a scholarship from Thailand Graduate Institute of Science and Technology (TGIST), NSTDA.

[1]   W. J. Koros, Ed, "*Barrier Polymers and Structures*", ACS Symposium Series; 423, American Chemical Society, Washington, DC, 1990, p.14.

[2]   W. E. Brown, "*Plastics in Food Packaging: Properties, Design, and Fabrication*", Dekker, New York, 1992, p.313.

[3]   S. Y. Lee, S. C. Kim, *Polym. Eng. Sci.* **1997**, *37*, 463.

[4]   M. J. Folkes, P.S. Hope, "*Polymer Blends and Alloys*", Blackie Academic, London, (1993).

[5]   W. Chinsirikul, T. C. Hsu, I. R. Harrison, *Polym. Eng. Sci.* **1996**, *36*, 2708.

[6]   O. Motta, L. D. Maio, L. Incarnato, D. Acierno, *Polymer* **1996**, *37*, 2373.

[7]   T. Nakinpong, "*A Study of Properties and Morphology of Low Density Polyethylene / Liquid Crystalline Polymer in-situ Composite Films*", Thesis, Mahidol University, 2000, p.57-60 and 70-73.

Macromol. Symp. **2004**, *216*, 281-292   281

Effects of Epoxidation Content of ENR on Morphology and Mechanical Properties of Natural Rubber Blended PVC

Jarunee Jeerupun,[1] *Jatuphorn Wootthikanokkhan,*[*][1] *Pranee Phinyocheep*[2]

[1] Division of Materials Technology, School of Energy and Materials, King Mongkut's University of Technology Thonburi, Bangkok 10140, Thailand
[2] Faculty of Science and Institute of Science and Technology for Research and Development, Mahidol University, Rama 6 Road, Bangkok 10400, Thailand

Summary: This research work has concerned a study on toughness of PVC/natural rubber (NR) blends compatibilized with epoxidized natural rubber (ENR). The aim of this work was to investigate the effect of degree of epoxidation on morphology and mechanical properties of the blends. Epoxidized natural rubber with a variety of epoxidation contents were prepared by reacting the NR latex with formic acid and hydrogen peroxide at various chemical contents. Chemical structure and epoxidation content of epoxidized natural rubber were evaluated by FTIR and ^{1}H-NMR techniques. After that, three grades of ENR with epoxidation contents of 15, 25 and 42 % (by mole) were further used for blending with PVC and NR in an internal mixer at 60 rpm and at 170 °C. From tensile and impact tests, it was found that tensile elongation and impact strength of the materials remarkably increased with degree of epoxidation. On the other hand, tensile strength and modulus of the materials rarely changed with the epoxidation content. An increase in toughness of the blends with epoxidation content was related to a better molecular interaction between PVC and ENR as suggested by torque-time curves of the materials.

Keywords: blends; compatibilization; poly(vinyl chloride) (PVC); rubber

Introduction

It has been known that poly(vinyl chloride) (PVC) is a notch sensitive plastic, having a low impact resistance. Various toughening agents have been compounded with PVC powders prior to a fabrication in order to improve toughness of the material. These toughening agents include methyl acrylate-butadiene-styrene (MBS), acrylonitrile-butadiene-styrene (ABS), and acrylate-methacrylate copolymer.[1] These toughening agents are phase separated from the PVC in a submicron level, capable of inducing crazing and shear yielding in the PVC matrix during a deformation process.[2]

Our interest in toughening of the PVC has concentrated on the use of natural rubber (NR) as an impact modifier. Because the NR is inexpensive and can be available in large quantities in south-east asia, it would be interesting and desirable if utilization and consumption of the NR could be extended. As mentioned earlier, in order to serve as a toughening agent, the rubber

© 2004 International Union of Pure and Applied Chemistry   DOI: 10.1002/masy.200451226

should be phase separated with PVC. This criterion could be achieved in our case owing to different polarity between PVC and the NR. However, compatibility between the two polymeric phases has yet to be improved. This is because of, again, the difference in polarity between the two polymers. In this regard, some kind of a compatibilizer should be used.

In this work, an epoxidized natural rubber (ENR) is considered to be a potential compatibilizer for this blending system. On one hand, chemical structure of the ENR is partially similar to that of a polyisoprene and/or the NR. On the other hand, epoxide groups in the ENR molecules could induce a high polarity nature of the rubber and thus promoting miscibility with PVC. In fact, extensive works on PVC-ENR blends [3-9] demonstrated that the two polymers are miscible, showing a hydrogen bonding peak in a FTIR spectrum as well as a single glass transition temperature in a DMTA thermogram. In relation to our work, it could be possible that the ENR would act as a compatibilizer, enhancing compatibility and adhesion between PVC and NR.

Unfortunately, little work concerning compatibilizing efficacy of ENR in PVC-NR system has been reported. Our previous study [10] demonstrated that the use of ENR could improve mechanical properties of PVC-NR blend. It was also found that the optimum blending composition with respect to toughness of the PVC/NR/ENR system is 80/10/10 weight ratio. However, an optimum degree of epoxidation of the ENR for this blend is still unknown and has to be clarified. Therefore, the aim of this work is to investigate the effects of epoxidation content of ENR on morphology and properties of the PVC/NR blends. Three types of ENR containing different epoxidation content were prepared and investigated.

Experimental

PVC resin (B0303 CLA, K-value = 58) was obtained from the Thai Plastic and Chemical Co. Ltd. Natural rubber (STR 5L) was supplied from the Thai Rubber Latex and Cooperation. Co., Ltd. High ammonia natural rubber latex (60 % d.r.c.) was supplied from the Rubber Research Institute of Thailand. Teric (16A16) used as a stabilizer for the latex was obtained from the East Asiatic (Thailand) Co., Ltd. Formic acid (98% w/w from BDH Ltd.) and hydrogen peroxide (from Merck Co., Ltd.) were analytical grade and were used as received.

The epoxidation was carried out in a 100 ml glass vessel at a temperature of 50 °C, using *in-situ* performic acid generated from formic acid and hydrogen peroxide.[11] Firstly, the 60 % dry rubber content (d.r.c.) latex was diluted to 30 % d.r.c., using distilled water. After that, 3 phr of Teric 16A16 was added to the diluted latex and then the mixture was stirred vigorously for 24 hrs in order to remove ammonia. Subsequently, a required amount of formic acid was added and the mixture was stirred for 10 min. Finally, a required amount of hydrogen peroxide was slowly added and the reaction was allowed to proceed for further 24 hrs. In this work, the ENR with a variety of epoxidation contents, specifically, 25, 35, and 50 % mol, were expected. This would be achieved by varying the amount of formic acid and hydrogen peroxide. After performing the reaction for a given time, the reacted products were separated from the reaction mixture by coagulating in methanol and then washed with distilled water before drying in a vacuum oven at 40 °C for 48 hrs.

Chemical structure of an epoxidized natural rubber (ENR) was confirmed by using Fourier transform infrared spectroscopy (FTIR) and proton nuclear magnetic resonance spectroscopy (^{1}H-NMR) techniques. Sample preparation for the FTIR experiment was conducted by casting a thin film of the material onto a ZnSe window cell. The experiment carried out by using a Bio-Rad (FTS 175) spectrophotometer. NMR experiment was carried out by using a Bruker (ADVANCE 300) spectrometer and the sample was prepared by dissolving about 15 mg of the ENR in 1 ml of CDCl$_3$. In order to determine an epoxidation content of an ENR, two characteristic ^{1}H-NMR peaks were considered in relation with the following equation;

$$\text{Epoxidation content (\% by mole)} = [A / (A + B)] \times 100 \tag{1}$$

where

A   =   Integrated area under the ^{1}H-NMR peak at about 2.70 ppm, representing the oxirane proton

B   =   Integrated area under the ^{1}H-NMR peak at about 5.14 ppm, representing an olefinic proton

PVC/NR/ENR blends with a weight composition of 80/10/10 were prepared by using an internal mixer (Haake Polylab Rheomex CTW 100, equipped with roller blade rotors). Blending process was carried out at a mixing temperature of 170 °C and at a rotor rotating speed of 60 rpm. In a typical blending process, PVC resin was firstly added to the mixer for 4 min in order to soften and melt the resin and then NR and ENR were concurrently added into the mixing chamber and the blending process was allowed to be continued for further 10 min. It is worth mentioning that no sign of a degradation of the PVC (such as a de-hydrochlorination) has been observed during the blending process. After blending, the material obtained from an internal mixer was crushed into small pieces by using a grinding machine. The grinded blend was then pass through a hot two roll mill operated at 190 °C and then fabricated in a compression mould. The molding process was carried out by using a LabTech

Engineering machine (L10-S-20) at 195 °C and at a clamping force of 17 ton. Mechanical properties of various blends were determined by tensile test and impact test. In a typical tensile test, dumbbell shaped specimens were prepared by stamping a sheet of polymer blend in accordance with an ASTM D638 (Type I). The test was carried out by using a Shimadzu AGS-500D at a crosshead speed of 50 mm/min, at a temperature of 25 °C. Specimens for the impact test were prepared in accordance with ASTM D256 to generate notched rectangular bars (63 mm ×12 mm × 3 mm) with 2.5 mm notch depth, 0.25 mm notch radius and at 45° notch angle. The impact test was carried out by using a Yasuda machine connected with a 4 J pendulum size.

Morphology of various blends was examined by using a scanning electron microscope (JEOL-JSM-5800) equipped with a backscattering electron detector. Specimen for the SEM experiment was prepared by cryo-microtoming the material under liquid nitrogen atmosphere. After that, the specimen was stained with OsO_4 vapor for 12 h before coating with gold using a sputtering unit. The SEM experiment was conducted by using an accelerating voltage of 15 kV.

Results and Discussion

Figure 1 shows an FTIR spectrum of the NR latex. Some characteristic absorption peaks, corresponding to functional groups in the natural rubber molecules can be noted, including the peaks at 1664 cm^{-1} (>C=CH-, stretching) and 836 cm^{-1} (=CH-, bending). This spectrum is significantly different from that of the product obtained from an epoxidation (Figure 2). From the Figure 2, it can be seen that new absorption peaks at about 874 cm^{-1} and 1250 cm^{-1} emerged. These are attributed to the presence of an epoxide ring or an oxirane ring in the ENR molecule. In addition, Figure 2 also shows a broad peak at about 3200-3500 cm^{-1}. This peak could only be observed when the epoxidation was carried out by using high amount of formic acid and hydrogen peroxide. This peak might be related to the presence of a hydroxyl group, which was obtained from a ring opening of the epoxy groups.

Results from ^{1}H-NMR spectrum of the product (Figure 3) further confirm the presence of epoxide ring in the product molecule. Peaks at about 2.7 ppm and 1.3 ppm could be observed and these represent the proton attached to the epoxide ring and the proton from a methyl group attached to the epoxide ring, respectively. Additionally, peaks at 3.4 and 3.9 ppm can also be observed and those could be attributed to some functional groups such as hydroxyl and hydrofuran, which were generated from a ring opening of the epoxy groups. [12-14] However, for the reaction carried out by using a relatively low formic acid and hydrogen peroxide content, those peaks were not detected.

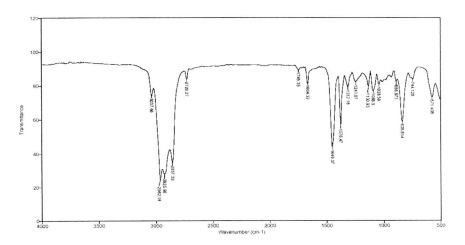

Fig. 1. FTIR spectrum of natural rubber.

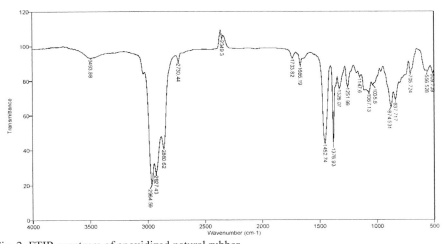

Fig. 2. FTIR spectrum of epoxidized natural rubber.

Table 1 shows different reaction parameters used and degree of epoxidation of an ENR determined by ¹H-NMR. Three types of ENR were successfully prepared i.e. ENR with epoxidation contents of 15, 25 and 42 %mol, that was controlled by varying the amount of formic acid and hydrogen peroxide, at a fixed temperature, time and the molar ratio of hydrogen peroxide to formic acid. It can be clearly seen that increasing the amount of formic acid and hydrogen peroxide resulted in increasing the degree of epoxidation. The results are in accordance with the work of Tanessriyanon et al.[15]

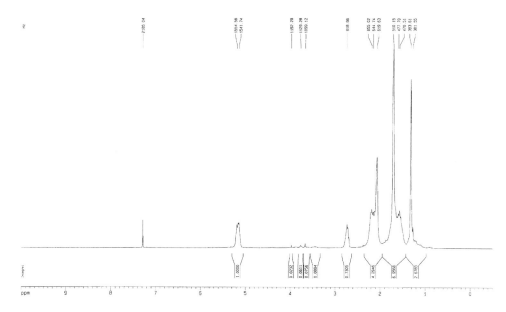

Fig. 3. ^{1}H-NMR spectrum of epoxidized natural rubber.

Table 1. Epoxidation content of various ENR as a function of reaction parameters.

Reaction No.	Formic acid Content (ml)	Hydrogen peroxide content (ml)	Epoxidation content (%)
1	12.6	111.1	15
2	17.6	155.6	25
3	20.9	221.9	42

Figure 4 shows overlaid torque-time profiles of various blends obtained from an internal mixer. The first peak at about 4 min. could be ascribed to a fusion peak of PVC resin whereas the latter peak could be attributed to a loading peak of both rubbers. After this peak, torque value gradually decreased with time and finally reaching equilibrium. It is interesting noting that an equilibrium torque of these materials changed with epoxidation content of the ENR in the blends. Generally speaking, the higher the epoxidation content, the higher the equilibrium torque value. In our opinion, this effect could be attributed to two main factors. Firstly, it might be possible that molecular interaction or miscibility between PVC and ENR increased, due to the fact that polarity of the ENR increased with the epoxidation content. In addition, it was also possible that viscosity of the ENR increased with the epoxidation content, leading to the higher mixing torque.

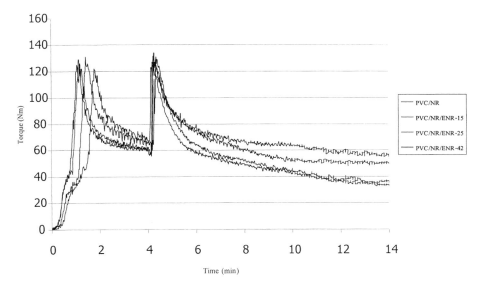

Fig. 4. Torque-time curves of various blends.

Figures 5-8 show SEM images of various blends. Dispersion of bright particles, representing the stained rubber phases, in PVC matrix (the darker background) could be observed. It is worth noting that some of the dispersed rubber phase has been etched, probably due to the cryo-microtoming process. Nevertheless, comparison between Figure 5 and 6 reveals that particle size of the NR phase significantly decreased after the ENR with 15 % epoxidation content was added. This result suggests that the ENR might acts as an emulsifier, reducing phase size of the NR phase. Furthermore, Figures 7-8 show that by blending with ENR containing higher degree of epoxidation (25 and 42 % mole), number of bright particles representing the rubber phases decreased. This result could be attributed to two main factors. Firstly, the higher the epoxidation content, the less the number of unsaturated bonds available for staining with OsO_4. In addition, it could be possible that the higher the degree of epoxidation, the greater the miscibility between ENR and PVC.

288

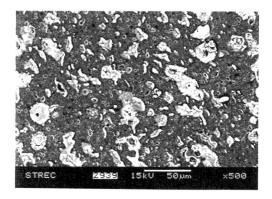

Fig. 5. SEM micrograph of PVC/NR blend.

Fig. 6. SEM micrograph of PVC/NR/ENR-15 blend.

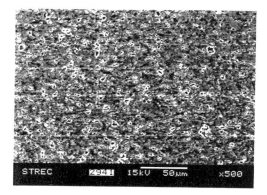

Fig. 7. SEM micrograph of PVC/NR/ENR-25 blend.

Fig. 8. SEM micrograph of PVC/NR/ENR-42 blend.

Changes in tensile properties of various blends as a function of epoxidation content could be seen from Figures 9-12. Ultimate tensile strength and modulus of these blends rarely changed with the epoxidation content. This might be related to the fact that weight composition between PVC and rubber phase is still 80/20 %, regardless of the epoxidation content. However, Figure 11 shows that tensile elongation of the blends significantly increased with epoxidation content. The above results imply that tensile toughness of the materials might be increased. In addition, result from the impact test (Figure 12) further confirms that impact strength of the blends increased with the epoxidation content. In theory,[15] impact resistance of a rubber toughened plastic could be related to morphology of the material. Particularly, particle size and ligament thickness are considered to be important factors, determining toughness of the material. For example, work by Liu et al. [16-17] on PVC/nitrile rubber blends shows that a critical ligament thickness of the blend corresponding to the transition from tough to brittle zones was 0.036 μm.

In relation to our work, however, addition of the ENR containing 15 % epoxidation content did not improve toughness of the blend even through the particle size has been remarkably decreased. In this regard, it seems that the effect of epoxidation content on impact properties of the PVC/NR blends might be attributed to some other factors, including miscibility between PVC and ENR phases. This statement has been supported by results from literature works [6] showed that by blending ENR with PVC, toughness of the materials could be improved even through the two phases are miscible. According to work by Ishiaku et al [6], FTIR spectrum of the PVC/ENR blend revealed an absorption peak representing a hydrogen bonding. In addition, some other relevance work [7] found a single T_g from a DMTA thermogram of the blends. In relation to toughness of the blends, it was believed that some of the ENR rubber phases were retained in a form of finely dispersed particles and cross-linked particles. In other words, the ENR rubber phases did not disappeared completely as is the case of a plasticizer. Therefore, the presence of these finely dispersed particles could contribute to an improved toughness of the blends. In relation to our work herein, it could be possible that an increase in miscibility between ENR and PVC lead to an increase in toughness of the blends.

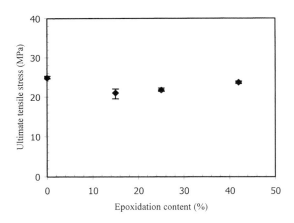

Fig. 9. Tensile strength of the PVC/NR/ENR blends as a function of epoxidation content.

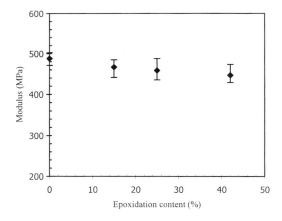

Fig. 10. Modulus of the PVC/NR/ENR blends as a function of epoxidation content.

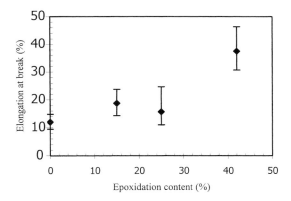

Fig. 11. Elongation of the PVC/NR/ENR blends as a function of epoxidation content.

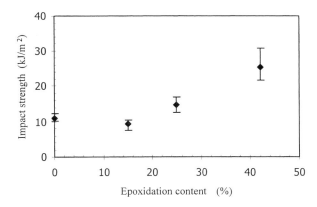

Fig. 12. Impact strength of the PVC/NR/ENR blends as a function of epoxidation content.

Conclusion

It has been shown that impact properties of PVC blended with natural rubber (NR) could be improved by adding an epoxidized natural rubber. It was also found that the greater the degree of epoxidation, the higher the toughness. The effect was related to an increase in mixing torque of these blends with the epoxidation content.

Acknowledgements

The authors are sincerely grateful to the Thai Plastic and Chemical Co. Ltd., for supplying the PVC resin used in this research.

292

[1]   J. C. Stevenson, *PMAD (Polymer Modifiers and Additives Division SPE Newsletter)*, **1994**, *12(2)*, 11.
[2]   L. H. Sperling, in: *"Polymeric Multicomponent Materials; An Introduction"*, John Wiley & Sons, Inc., Canada, **1997**, p. 243.
[3]   Z. A. Nasir, U. S. Ishiaku, Z. A. Mohd Ishak, *Journal of Applied Polymer Science* **1993**, *47*, 951.
[4]   Z. A. Nasir, C. T. Ratnam, *Journal of Applied Polymer Science* **1989**, *38*, 1219.
[5]   A. Mousa, U. S. Ishiaku, Z. A. Mohd Ishak, *Plastics, Rubber and Composites Processing and Applications* **1997**, *26(8)*, 331.
[6]   U. S. Ishiaku, M. Nasir, Z. A. Mohd Ishak, *Journal of Vinyl Technology* **1994**, *16(4)*, 219.
[7]   U. S. Ishiaku, M. Nasir, Z. A. Mohd Ishak, *Journal of Vinyl Technology* **1994**, *16(4)*, 226.
[8]   U. S. Ishiaku, M. Nasir, Z. A. Mohd Ishak, *Journal of Vinyl & Additive Technology* **1995**, *1(3)*, 142.
[9]   U. S. Ishiaku, M. Nasir, Z. A. Mohd Ishak, *Journal of Vinyl & Additive Technology* **1995**, *1(2)*, 66.
[10]   W. Sukkaew, M.Eng. Thesis, King Mongkut's University of Technology Thonburi, **2000**.
[11]   I. R. Gelling, *Journal of Natural Rubber Research* **1991**, *6(3)*, 184.
[12]   M. C. S. Perera, J. A. Elix, J. H. Bradbury, *Journal of Polymer Science: Part A: Polymer Chemistry* **1998**, *26*, 637.
[13]   S. Roy, B. R. Gupta, B. R. Maiti, *Journal of Elastomers & Plastics* **1990**, *22*, 280.
[14]   S. Roy, C. S. S. Namboodri, B. R. Maiti, B. R. Gupta, *Polymer Engineering & Science* **1993**, *33(2)*, 92.
[15]   Y. Tanassreeyanon, P. Phinyocheep, M. Sea-Dan and N. Ratanasom, *9th International Seminar on Elastomers*, Kyoto, Japan, April 2-4, **2003**, 1A-27, 21.
[16]   A. A. Collyer, in: *"Rubber Toughened Engineering Plastics"*, Chapman & Hall, London, **1994**, p. 37.
[17]   Z. H. Liu, X. D. Zhang, X.G. Zhu, R. K. Y. Li, F. S. Wang, C. L. Choy, *Polymer* **1998**, *39(21)*, 5019.
[18]   Z. H. Liu, X. D. Zhang, X.G. Zhu, Z. N. Qi, F. S. Wang, R. K. Y. Li, C. L. Choy, *Polymer* **1998**, *39(21)*, 5019.

Effects of Solution Concentration, Emitting Electrode Polarity, Solvent Type, and Salt Addition on Electrospun Polyamide-6 Fibers: A Preliminary Report

Chidchanok Mit-uppatham, Manit Nithitanakul, Pitt Supaphol*

The Petroleum and Petrochemical College, Chulalongkorn University, Soi Chula 12, Phyathai Road, Wang-Mai, Pathumwan, Bangkok 10330, Thailand
E-mail: pitt.s@chula.ac.th

Summary: Electrospinning is a process by which ultrafine fibers which have diameters in the range of tens of nanometers to less than ten of micrometers can be produced. This process utilizes expulsion of charges as a means to very thin fiber formation. In this short report, the effects of some of the influencing solution and process parameters (i.e. solution concentration, emitting electrode polarity, solvent type, and salt addition) on morphological appearance of electrospun polyamide-6 fibers were investigated based on visual observation of a series of scanning electron micrographs. It was found that all of the parameters studied played important roles in determining morphology and sizes of the fibers obtained.

Keywords: electrostatic spinning; polyamide-6; ultrafine fibers

Introduction

Electrostatic spinning or electrospinning is a newly-arrived fiber spinning process for producing ultra-fine fibers with average diameters in the range of tens of nanometers to sub-micrometers. Naturally, the ultra-fine fibers from this process are obtained as a non-woven fabric, which exhibits several interesting characteristics, such as small pore sizes between adjacent fibers, high porosity, and high specific surface area. These characteristics can be of tremendous uses in some applications.[1]

The basic principles of the electrospinning process are concerned with the application of a high electrical potential to a polymer solution or melt across a finite distance between a nozzle and a collective target. The polarity of the emitting electrode (i.e. the one that is in contact with the polymer solution or melt) can be either positive or negative. When an electrostatic field is applied, charges are built up on surface of a droplet of the polymer solution or melt at the tip of the nozzle. The charges destabilize the hemispherical shape of the droplet into a cone shape at a critical value of the applied electrostatic field. With further

DOI: 10.1002/masy.200451227

increase in the applied electrostatic field, the electrostatic force overcomes the surface tension, causing a charged stream of polymer solution or melt (i.e. a charged jet) to be ejected from the tip of the cone. The charged jet travels in a straight line for a few centimeters as its diameter thins down appreciably, before undergoing a bending instability during which the diameter of the jet continues to decrease tremendously. Finally, fibers are collected on a grounded collector plate.[2]

In the electrospinning from a polymer solution, various solution properties are affecting the morphology of the obtained fibers. Some of these are solution concentration, viscosity, surface tension, and conductivity. D eitzel and co-workers[3] found that, in order to obtain poly(ethylene oxide) (PEO) fibers of good appearance, the viscosity of the PEO solution should be in the range of 1 to 20 poise and the surface tension in the range of 35 to 55 dyne/cm. At viscosity values greater than 20 poise, electrospinning became prohibitive due to the extremely high cohesiveness of the solutions which restricts the flow; whereas, at viscosity values lower than 1 poise, droplets were abundant due to the extremely low cohesiveness of the solutions which causes the jet to break up.[3] Baumgarten[4] found that uniform acrylic fibers were obtained when the viscosity of the stock solutions was in the range of 1.7 to 215 poise. He also found that addition of small amount of inorganic salts into the solutions helped promote the fiber formation with no beads present, a result of the increased charge density in a jet segment. The fiber diameters were also found to decrease with addition of the salts.[5,6]

In the present contribution, the effects of some of the influencing solution and process parameters (i.e. solution concentration, emitting electrode polarity, solvent type, and salt addition) on morphological appearance of electrospun polyamide-6 fibers are preliminarily reported.

Experimental Details

Polyamide-6 resin ($\overline{M}_w$ = 20,000 Da) was courteously supplied by Asia Fiber Public Co., Ltd. (Thailand). In order to elucidate the effect of solution viscosity on morphology of the obtained fibers, solutions of polyamide-6 in 85% v/v formic acid (Carlo Erba) were prepared in various concentrations, ranging from 10 to 46 percent by weight (wt.%). The effect of solvent type was investigated by mixing m-cresol with formic acid in various ratios, while sodium chloride (NaCl) was added in solutions of polyamide-6 in formic acid in order to observe the effect of salt addition. Each stock solution prepared was characterized for viscosity, surface tension, and conductivity using a Brookfield DV-III programmable rheometer, a KrÜss K10T tensiometer, and a Orion 160 conductivity meter, respectively.

Each as-prepared polyamide-6 solution was stocked in a 50-ml glass syringe before electrospinning. A stainless steel needle of gauge no. 26 was used as the nozzle. Feed rate of the solution was controlled by fixing the flow rate of nitrogen gas, connected to the feed side of the glass syringe. A piece of aluminum sheet was used as the grounded collective screen. A Gamma High Voltage Research D-ES30PN/M692 DC power supply (Florida, USA) was connected to the nozzle and the grounded collective screen. The polarity of the emitting electrode (e.g. the one connecting to the nozzle) could either be positive or negative. In this particular report, the nozzle-to-collector distance of 10 cm and the applied DC potential (either positive or negative polarity) of 21 kV were fixed.

The morphological appearance of the obtained polyamide-6 fibers was observed by a JEOL JSM-5200 scanning electron microscope.

Results and Discussion

It has been shown to some extent that solution properties (i.e. viscosity, surface tension, and conductivity) play an important role in the morphological appearance of the obtained electrospun polymeric fibers. In order to elucidate such a statement to the case of electrospun polyamide-6 fibers, polyamide-6 solutions in 85% v/v formic acid with the concentration being in the range of 10 to 46 wt.% were characterized for their viscosity, surface tension, and conductivity values. It was found that the viscosity value of the solutions was found to tremendously increase from 40 to 4058 cp with increasing polyamide-6 concentration from 10 to 46 wt.%, while the surface tension and conductivity values were found to increase very slightly.

Qualitatively, the results showed that when the viscosity value of the polyamide-6 solutions was lower than 135 cp (corresponding to the concentration of the solution of 16 wt.%), only droplets were present. The formation of the droplets for solutions having viscosity values lower than 135 cp can be described based on the analysis of forces acting on a small segment of a charged jet. In this case, six types of forces can be considered: they are 1) body or gravitational force, 2) electrostatic force which carries the charged jet from the nozzle to the collective target, 3) Coulombic force which tries to push apart adjacent charged carriers within the jet segment and is responsible for the stretching of the charged jet during its flight to the target, 4) viscoelastic force which tries to prevent the charged jet from stretching, 5) surface tension which prevents the surface of the charged jet from stretching, and 6) drag force from the friction between the charged jet and the surrounding air.[7]

To explain the occurrence of the droplets for solutions having viscosity value lower than 135 cp can be described by considering the interplay between three most important forces being responsible for the formation of elongated jet, i.e. Coulombic force, viscoelastic force, and surface tension. At low viscosity values, the viscoelastic force was comparatively smaller than the Coulombic force. This resulted in the over-stretching of a charged jet, hence the break-up of the charged jet into many small spherical droplets as a result of the surface tension. On the contrary, for solutions of higher viscosity, the viscoelastic force became larger in comparison with the Coulombic force (due mainly to the increased number of chain entanglements). The increase in the viscoelastic force was sufficient to prevent a charged jet from breaking up into small droplets and to allow the electrostatic stress to further elongate the jet which finally thins down the diameter of the jet tremendously.

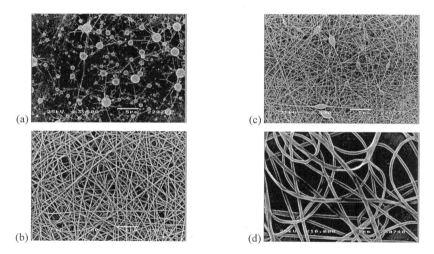

Figure 1. Electrospun products from solutions of polyamide-6 in formic acid as a function of solution concentration (solution viscosity) with the positive polarity of the emitting electrode for (a) 16 wt.% (135 cp), (b) 28 wt.% (689 cp), and (c) 40 wt.% (2445 cp) and with the negative polarity of the emitting electrode for (d) 38 wt.% (1928 cp).

Figure 1 shows selected scanning electron micrographs of products obtained from the electrospinning of polyamide-6 solutions in formic acid for four different concentrations (and hence four different viscosity values) using either positive or negative polarity of the emitting electrode. Evidently, droplets were more prevalent when the viscosity of the polyamide-6 solution was 135 cp (see Figure 1a). With an increase in the viscosity of the solution to 689 cp, fibers with spindle-like droplets were present along with pure fibers (see Figure 1b). With further increase in the viscosity of the solution to 2445 cp, only uniform fibers were obtained (see Figure 1c). Qualitatively, diameters of the obtained fibers were found to increase with increasing viscosity of the solutions.

What has been presented in Figure 1a to 1c was obtained from spinning polyamide-6 solutions with the polarity of the emitting electrode being positive. Interestingly, when the polarity of the emitting electrode was negative, flat-shaped fibers were instead observed, particularly when the viscosity of the solutions exceeded 1928 cp (see Figure 1d). In addition to the formation of flat fibers, it is observed that the negative polarity of the emitting electrode resulted in fibers with much larger diameters than the positive one. The reasons for the larger diameters of the obtained fibers when the negative polarity was used might be an increase in the mass flow rate (a result of the higher charge density) and for the formation of flat fibers might be the collapsing of the dried skin layer of a jet after the solvent inside the jet had evaporated.[8]

| (a) | (b) | (c) |

Figure 2. Electrospun products from solutions of 32 wt.% of polyamide-6 in a) *m*-cresol and mixed solvents of formic acid and *m*-cresol having the compositional ratio between formic acid and *m*-cresol of b) 80:20 and c) 60:40 v/v with positive polarity of the emitting electrode.

The effect of solvent type on morphological appearance of the obtained polyamide-6 fiber was investigated by dissolving polyamide-6 in *m*-cresol or mixed solvents of formic acid and *m*-cresol having the compositional ratio between formic acid and *m*-cresol of 90:10, 80:20, 70:30, 60:40, and 50:50 v/v prior to electrospinning. The viscosity of the mixed solvents was found to increase, while the conductivity was found to decrease, with increasing amount of added *m*-cresol. Figure 2 illustrated selected results of fibers obtained from solutions of polyamide-6 in *m*-cresol and the mixed solvents of 80:20 and 60:40 compositional ratios.

According to Figure 2a, electrospun fibers were not observed when pure *m*-cresol was used as the solvent. The most likely explanation may be due to the higher viscosity and lower conductivity of the resulting solution and to the higher boiling point of *m*-cresol (i.e. 203°C) in comparison with that of formic acid (i.e. 118°C). Lee and co-workers[9] found that both dielectric constant and conductivity of the spinning solution were key factors signifying the electrospinning process.

Fibers of smooth surface were observed with the solutions having *m*-cresol content between 10 to 30 percent by volume (i.e. vol.%) (see, for example, Figure 2b), while fused fibers were observed with the solutions having *m*-cresol content greater than 40 vol.% (see, for example, Figure 2c). The higher boiling point of *m*-cresol should be responsible for the formation of fused fibers observed from solutions of polyamide-6 in mixed solvents having high *m*-cresol content. Additionally, diameters of the fibers were found to increase with increasing amount of *m*-cresol in the solutions, likely a result of the increased viscosity and the decreased conductivity values.

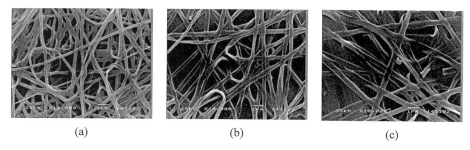

(a)   (b)   (c)

Figure 3. Electrospun products from solutions of 32 wt.% of polyamide-6 in formic acid using negative polarity of the emitting electrode with a) 2, b) 3, and c) 4 percent by weight of NaCl salt addition.

The effect of solution conductivity on morphological appearance of the obtained polyamide-6 fibers was also investigated. The solution conductivity was varied by varying the content of NaCl salt in the system. The concentration of NaCl was varied in the range of 1 to 5 wt.%. Note that the low concentrations of NaCl salt added were to ensure complete solubility of the salt in the solutions. The conductivity of the resulting solutions was found to increase with increasing amount of NaCl salt added. Figure 3 shows morphological appearance of the obtained polyamide-6 fibers as a result of salt addition at different amounts using negative polarity of the emitting electrode. Clearly, the fibers obtained were flat and, with increasing salt content, the size of the fibers was found to increase. It is hypothesized that both the application of the negative polarity and the addition of NaCl salt helped increase the mass flow rate, hence an increase in the size of the fibers obtained with the flat shape being formed by the much slower evaporation rates as a result of the larger size of the fibers.

Conclusions

In this short report, the electrospinning technique was used to produce ultra-fine polyamide-6 fibers. The effects of some of the influencing solution and process parameters (i.e. solution concentration, emitting electrode polarity, solvent type, and salt addition) on morphological appearance of the obtained fibers were visually observed from a series of scanning electron micrographs. It was found that solutions with high enough viscosity values were necessary to result in electrospun polyamide-6 fibers having uniform diameters. Mixing *m*-cresol with formic acid to be used as the mixed solvent for dissolving polyamide-6 affected the morphological appearance a great deal. It was found that diameters of the fibers obtained increased with increasing amount of *m*-cresol and, at *m*-cresol content of greater than 40 percent by weight, fused fibers were formed, attributable to the higher boiling of *m*-cresol. It was found that flat fibers were obtained with use of the negative polarity of the emitting electrode. Lastly, increased conductivity of the solutions by increasing amount of NaCl salt addition resulted in fibers of larger sizes.

Acknowledgments

This work was financially supported by an invention grant and a research grant from the Rachadapisek Somphot Endowment Fund, Chulalongkorn University. Partial support from the Petroleum and Petrochemical College (PPC) and the Petroleum and Petrochemical Technology Consortium is gratefully acknowledged. Helpful discussion with Dr. Ratthapol Rangkupan of the Metallurgy and Materials Science Research Institute (MMRI), Chulalongkorn University is gratefully acknowledged.

[1]   J. Doshi, D. H. Reneker, *J. Electrostatics* **1995**, *35*, 151.
[2]   D. H. Reneker, A. L. Yarin, H. Fong, S. Koombhongse, *J. Appl. Phys.* **2000**, *87*, 4531.
[3]   J. M. Deitzel, J. Kleinmeyer, D. Harris, N. C. Beck Tan, *Polymer* **2001**, *42*, 261.
[4]   P. K. Baumgarten, *J. Colloid Interface Sci.* **1971**, *36*, 71.
[5]   H. Fong, I. Chun, D. H. Reneker, *Polymer* **1999**, *40*, 4585.
[6]   X. Zong, K. Kim, D. Fang, S. Ran, B. S. Hsiao, B. Chu, *Polymer* **2002**, *43*, 4403.
[7]   L. Wannatong, A. Sirivat, P. Supaphol, *Polym. Int.*, accepted for publication.
[8]   S. Koombhongse, W. Liu, D. H. Reneker, *J. Polym. Sci. B: Polym. Phys.* **2001**, *39*, 2598.
[9]   K. H. Lee, H. Y. Kim, M. S. Khil, Y. M. Ra, D. R. Lee, *Polymer* **2003**, *44*, 1287.

Electrospun Non-Woven Fabrics of Poly(ε-caprolactone) and Their Biodegradation by Pure Cultures of Soil Filamentous Fungi

Kousaku Ohkawa,[1] *Hakyong Kim,*[2] *Keunhyung Lee,*[3] *Hiroyuki Yamamoto*[1]

[1] Institute of High Polymer Research, Faculty of Textile Science and Technology, Shinshu University, Ueda 386-8567, Japan
E-mail: kohkawa@giptc.shinshu-u.ac.jp
[2] Department of Textile Engineering, Chonbuk National University, Chonju 561-756, Republic of Korea
[3] Department of Advanced Organic Materials Engineering, Chonbuk National University, Chonju 561-756, Republic of Korea

Summary: The biodegradation of electrospun nano-fibers of poly(ε-caprolactone) was firstly investigated, using pure-cultured soil filamentous fungi. In the biochemical oxygen demand test, the biodegradation of the nano-fiber exceeded 20–30% carbon dioxide generation. The biodegradabilities decrease with increase of the mean fiber diameter.

Keywords: biodegradation; electrospinning; nano-fibers; poly(ε-caprolactone); soil fungi

Introduction

Electrospinning is a unique technique able to prepare the polymer fibers having diameters in the ranges from nano- to a few micrometers.[1] Non-woven fabrics composed of electrospun nano-fibers have a large surface area per unit mass and a small pore size.[2,3] Biodegradable non-woven fabrics have been recently focused because of their wide ranges of applications.

Poly(ε-caprolactone) (PCL) is one of the most important class of biodegradable polymer which exhibits a high-level biodegradability in almost all given situations, considering *in vitro*, *in vivo*, and environmental usage.[4,5] As for the electropsun non-woven fabric of PCL, its biodegradation property has been less known. We report here the preparation of the electrospun non-woven fabrics of PCL and their biodegradabilities with respects to environmental usage.

The purposes of this study are as follows; (i) The first evaluation of the environmental biodegradability of electrospun PCL nano-fibers; (ii) Laboratory experiments of PCL biodegradation using the single strained soil filamentous fungi; (iii) Relationship between the structural characteristics and the biodegradabilities of PCL nano-fibers.

Experimental

Materials

PCL (M_n 80,000) was purchased from Aldrich (Milwaukee, USA). This material was dissolved in a mixture of methylene chloride and N,N-dimethylformamide (85/15, vol.-%) at room temperature. The viscosity of PCL solution was determined by using the rheometer (DV III, Brookfield Co., USA) equipped with a spindle No. 3 at 100 rpm at 20°C.

Electrospinning of Poly(ε-caprolactone)

PCL solution prepared was poured in a 5-mL syringe attached to a capillary tip of about 1 mm diameter. The copper wire connected to an anode was inserted into the solution and a cathode was attached to a grounded rotating metallic collector. The electric field was produced by a high-voltage power supply (CPS-60 K02v1, Chungpa EMT, Co., Republic of Korea).

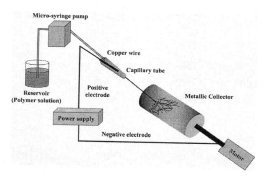

Figure 1. Electrospinning apparatus.

PCL non-woven fabrics were made from solutions with viscosities of 150, 210, and 310 centipoise under identical conditions with 15 kV (applied voltage) and 120 mm (distance between capillary tip and collector). After electrospinning, non-woven fabrics produced were dried in a vacuum oven for 1 week to remove the residual solvent.

Microscopic Observation of Fungal Growth

Seven species of single strained soil filamentous fungi, *Aspergillus oryzae*, *Penicillium caseicolum*, *P. citrinum*, *Mucor* sp., *Rhizopus* sp., *Curvularia* sp., and *Cladosporium* sp. were employed, according to our previous article.[6] The non-woven PCL sheet samples were cut into pieces (10×10 mm^2). Three pieces were placed in a dish and then the modified Czapeck medium (1/100-reduced carbon source) and the sporangia suspensions were then added at 25°C. The growth of the microorganisms and the collapse of the nano-fiber accompanying the biodegradation were observed using a scanning electron microscope.

Quantification of Biodegradability

The biochemical oxygen demand (BOD)-biodegradation test was conducted using the method reported by Doi et al.[7] The non-woven PCL samples (7–8 mg) were placed in a BOD reactor, and the modified Czapeck medium and sporangia suspensions of the fungi were then added.

The biodegradation test was carried out for 30 days, and the BOD was continuously measured as the CO_2 evolution volume. The BOD-biodegradation of the fiber sample was calculated using an equation: Degradation (%) = $100 \times (BOD^t - BOD^b) / TOD$, where BOD^t and BOD^b are values of the fiber sample and blank test, respectively. The TOD is the theoretical oxygen demand.

Results and Discussion

Characterization of Non-Woven Fabrics

Table 1 represented the distribution of fiber diameters in the electrospun non-woven fabrics of PCL. Three kinds of the non-woven fabrics having nano-scale fibers were prepared from the PCL solutions at three different viscosities, 150, 210, and 310 cPs. At the solution viscosity of 150 cPs, the peak fraction in the diameter distribution located around 200–400 nm, and the frequency distribution was approximately 70 %. The mean fiber diameter in the electrospinning at 150 cPs was 330 nm. SEM photographs visually proved that the thicker fibers were prepared in the higher viscosity. The estimated mean fiber diameters obtained in the 210 and 310 cPs conditions were 360 and 510 nm, respectively. The sample names were designated as the PCL330, PCL360, and PCL510, from their mean diameters.

Table 1. Characterization of electrospun PCL nan-woven fabrics

Solution viscosity	150 cPs	210 cPs	310 cPs
Mean diameter	330 nm	360 nm	510 nm

Fungal Growth on PCL Nano-Fibers

The PCL non-woven fabrics were subjected to microbial degradation in a culture with seven species of soil fungi. Among the fungi tested, the *Mucor* sp. and *Rhizopus* sp. remarkably degraded the PCL non-woven fabrics, and the process of the degradation could be observed under a scanning electron microscope. Figure 2 shows the SEM images during the growth of *Mucor* sp. on the PCL360 sample. Before biodegradation (Figure 2a) , the non-woven fabrics exhibited the fibril structure with nano-scale diameter. In the initial phase, the filament of *Mucor* sp. became being growing on the surface of the PCL nano-fibers. After a 7-days cultivation, progress in surface coverage by the fungal body is clearly observed. The secretes from the fungal body covered the surface and digested the nano-fibers.

Further cultivation after 14 days confirmed a complete coverage of the nano-fiber surface by the fungal filaments. The maturation of the *Mucor* sp. was finally represented by the development of fungal spores (Figure 3b). These results indicated that the life cycle of *Mucor* sp. could be run during the cultivation on the PCL nano-fiber as well as that the electrospun PCL non-woven fabrics could be the nutrition for the *Mucor* sp. The metabolism of the PCL digests in the fungal body was subsequently examined by the BOD tests.

Table 2. Biodegradabilities of PCL non-woven fabrics by soil filamentous fungi

Filamentous fungi	CO_2 production[a] (micro mol)			Biodegradation (%)		
	PCL510	PCL360	PCL330	PCL510	PCL360	PCL330
Rhizopus sp.	64	75	116	17	20	31
Mucor sp.	45	53	75	12	14	20
Cladosporium sp.	42	46	60	11	12	16
Aspergillus oryzae	14	14	25	4	4	7
Curvularia sp.	14	10	14	4	3	4
Penicillum caseicolum	12	9	12	3	2	3
Penicillium citrinum	4	4	6	1	1	2

a) data after a 28-days cultivation.

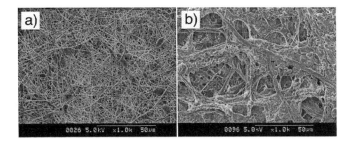

Figure 2. SEM observation of biodegradation process of PCL non-woven fabrics by *Mucor* sp.: (a), before degradation; (b), after incubation for 14 days.

Effect of PCL Nano-Fiber Diameters on Biodegradation by Soil Fungi

In the BOD tests, four species of fungi, *Rhizopus* sp, *Mucor* sp, *A. oryzae*, and *Cladosporium* sp. degraded all three kinds of PCL samples, producing the CO_2 gas. To summarize the results from the BOD tests, the relationship between the mean diameter of the PCL nano-fiber and the biodegradability was examined (Table2). In the all four fungi concerned, the degree of biodegradation is significantly increased when the mean fiber diameter below 360 nm. The biodegradability of the PCL non-woven fabrics will be constant above the mean diameter of 510 nm. This result suggests that the biodegradability of the non-woven fabrics of PCL might be regulated by the fiber diameter lower than 360 nm.

Conclusion

In the present study, we firstly performed the laboratory experiments on the biodegradability of electrospun non-woven fabrics of PCL using seven species of soil filamentous fungi. The following findings were obtained; (i) Single strained soil fungi were grown on the PCL non-woven fabrics, degrading the PCL nano-fiber and utilizing them as the nutrition, and (ii) The biodegradability of the PCL non-woven fabrics depends on their nano-fiber diameters. This finding could be expand to general electrospinnig system dealing with polyester-based materials, including the polymers of L-lactate and β-hydoxybutylate, which are oriented to the environmental uses.

Acknowledgements

This work was partly supported by Grants-in-aid for The 21st Century COE Program and for Scientific Research (No. 13556057 and No. 13555178) by the Ministry of Education, Culture, Sports, Science, and Technology of Japan.

[1]   J. Doshi, D. H. Reneker, *J. Electrostat.* **1995**, *35*, 151.
[2]   K. H. Lee, H. Y. Kim, Y. M. La, D. R. Lee, N. H. Sung, *J. Polym. Sci.* **2002**, *40*, 2259.
[3]   K. H. Lee, H. Y. Kim, Y. J. Ryu, K. W. Kim, S. W. Choi, *J. Polym. Sci.* **2003**, *41*, 1256.
[4]   D. Goldberg, *J. Environ. Polym. Degrad.* **1995**, *3*, 61.
[5]   A. C. Albertsson, R. Renstad, B. Erlandsson, C. Eldsater, S. Karlsson, *J. Appl. Polym. Sci.* **1998**, 70, 61.
[6]   K. Ohkawa, M. Yamada, A. Nishida, N. Nishi, H. Yamamoto, *J. Polym. Environ.* **2000**, *8*, 59.
[7]   Y. Doi, K. Kasuya, H. Abe, N. Koyama, S. Ishiwatari, K. Takagi, Y. Yoshida, *Polym. Degrad. Stab.* **1996**, *51*, 281.